THE VIC
FERN C

Sarah Whittingham

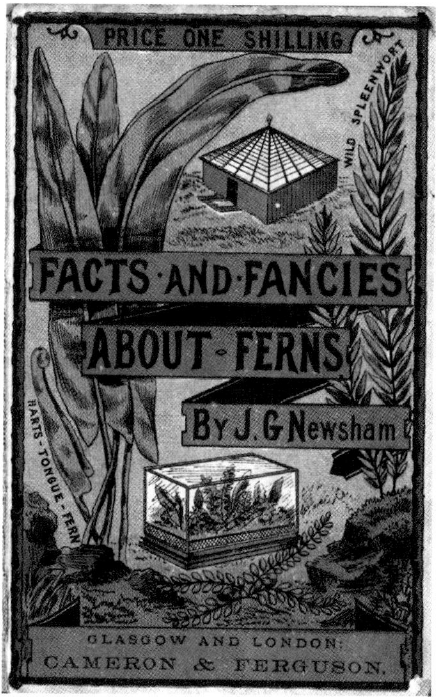

SHIRE PUBLICATIONS

Published in Great Britain in 2011 by Shire Publications Ltd, Midland House, West Way, Botley, Oxford OX2 0PH, United Kingdom.
44-02 23rd St, Suite 219, Long Island City, NY 11101, USA.

E-mail: shire@shirebooks.co.uk www.shirebooks.co.uk

A CIP catalogue record for this book is available from the British Library.

Shire Library no. 571 • ISBN: 978 0 74780 746 9

Sarah Whittingham has asserted her right under the Copyright, Designs and Patents Act, 1988, to be identified as the author of this book.

Designed by Ken Vail Graphic Design, Cambridge, UK and typeset in Perpetua and Gill Sans.
Printed in China through Worldprint Ltd.

11 12 13 14 15 10 9 8 7 6 5 4 3 2

COVER IMAGE
The Kibble Palace, Glasgow Botanic Gardens, photographed in 1997 before restoration.

TITLE PAGE IMAGE
Facts and Fancies About Ferns. Forming a Cheap and Complete Hand-Book for the Amateur Fern-Grower, Comprising the Various Modes of Culture &c, by John George Newsham (c. 1876).

CONTENTS PAGE IMAGE
Illustration of the maidenhair fern (*Adiantum capillus-veneris*) from Anne Pratt, *The Ferns of Great Britain and Their Allies: The Club-mosses, Pepperworts, and Horsetails* (1855).

ACKNOWLEDGEMENTS
For Sue and Matthew

Many thanks to the British Pteridological Society for a research grant from their Centenary Fund. Also, thank you to all those – alas too many to name here – who have given me information about all sorts of fern-related matters.

Illustrations were supplied by and are reproduced by kind permission of the following:

The Bodleian Library, University of Oxford (per.256 c.4, p. 389), page 23 (bottom); The British Pteridological Society, pages 19 and 20; Crown Copyright: Royal Commission on the Ancient and Historical Monuments of Scotland, page 56; Culture and Sport Glasgow (Museums), page 25 (bottom); The Eden Project/Simon Burt, Apex Photo Agency, page 57; John Harris, Dewstow Gardens, page 30 (top); Shrewsbury Museums Service, page 44 (top right and bottom left); Trewidden Garden/©Keith Hargreaves, page 33 (top).

All other illustrations are from the author's collection, of which the following are reproduced by kind permission of:

Bristol Central Library, page 14 (bottom); Chelsea Physic Garden, pages 8 and 36 (top); Cheshire East Council, who manage and finance Tatton Park on behalf of the National Trust, page 32; Wallace and Katherine Fyfe, page 28 (top and bottom); Glasgow Botanic Gardens, front cover; Lynne and Peter MacLean, page 52; The National Trust, page 30 (bottom); Nottingham City Council, page 25 (top); Oxford University Museum of Natural History, page 44 (top left); Sefton Council, page 36 (bottom); The Shuttleworth Trust, page 29 (bottom); Vale of Glamorgan Council, page 33 (bottom).

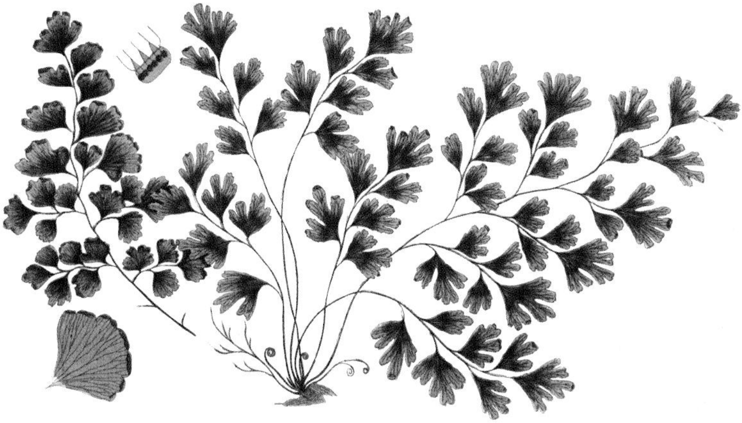

CONTENTS

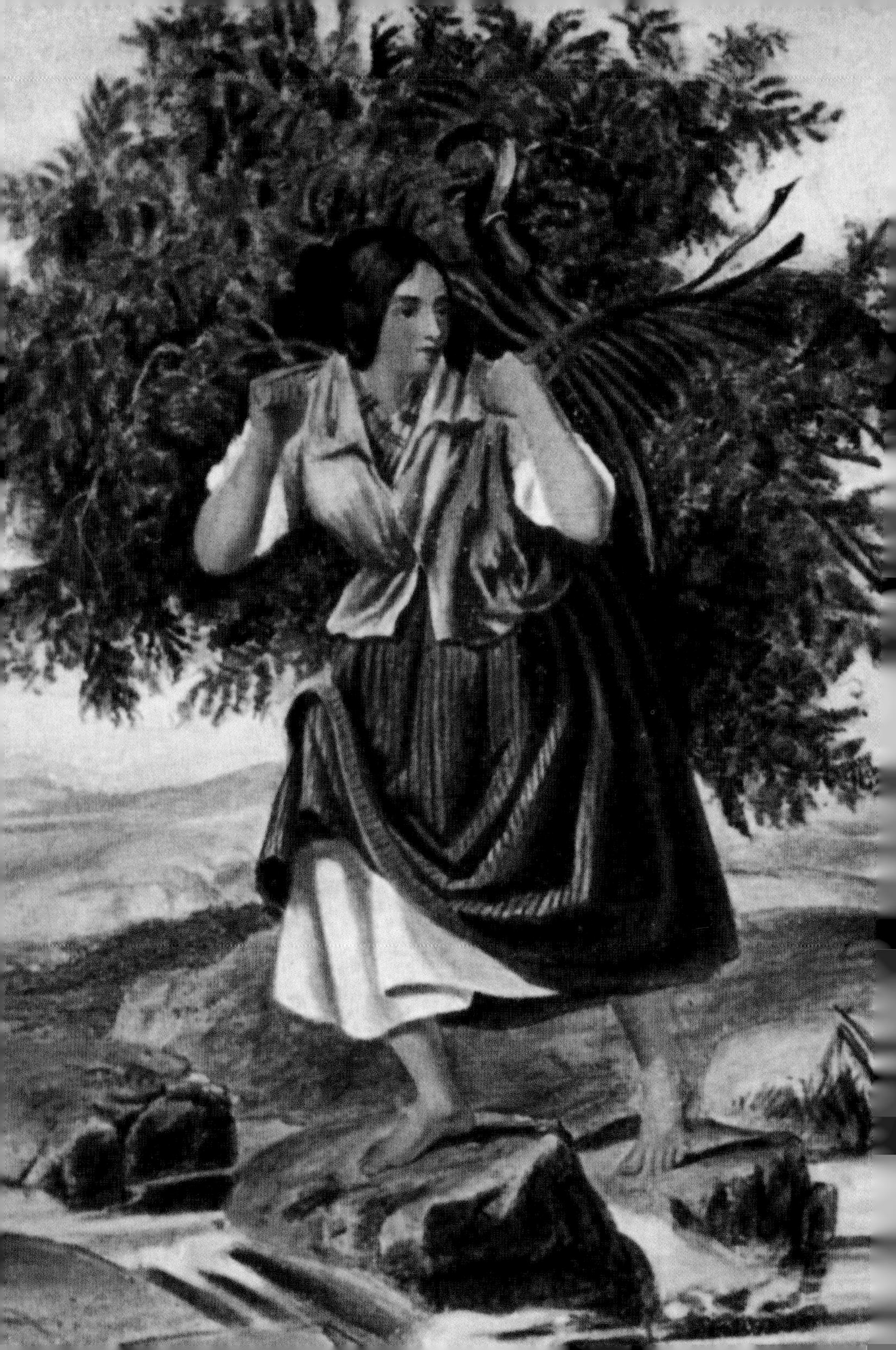

INTRODUCTION

FOR HUNDREDS of years ferns were regarded as magical and mysterious plants. This was largely because they reproduce in a rather unusual way; they do not have flowers or seeds, but produce microscopic spores, which germinate into prothalli that produce the gametes that form new ferns. Before this was discovered many myths grew up about the plant and the elusive 'fern seed', which was commonly believed to render its possessor invisible. Ferns have also long been used for medicinal purposes; from the Greeks onwards extracts were prescribed as a treatment for ailments as wide-ranging as asthma, hair loss, kidney complaints, and worms. The most common practical use of the fern – in the form of bracken or brake – was as litter for livestock and thatch for the roofs of cottages. For centuries, bracken ashes were also used in the production of lye soap and glass.

However, before the nineteenth century there was virtually no interest in growing native ferns for pleasure, and very few collections of exotic ferns in Britain. In 1689 the amateur botanist Sir Arthur Rawdon (1662–95) sent the gardener James Harlow to Jamaica to collect plants for his hot house at Moira in County Down. Harlow successfully brought back around 1,000 tropical plants, including the first exotic species of fern to be cultivated in Ireland. But at the beginning of the eighteenth century there were still very few exotic ferns in the United Kingdom. This was partly because if a fern died, it could only be replaced by the costly and difficult introduction of a new plant from abroad. Between 1770 and 1790, sixty-eight species of fern were imported, most of which went into private collections, and the rest to the Botanic Gardens at Kew.

A major breakthrough in the cultivation of foreign ferns was the discovery of a reliable method of raising them from spores. John Lindsay, a surgeon in Jamaica, reported his successful experiments in this field in 1794 (although he believed that the spores were seeds, and the plant's reproductive cycle wasn't fully comprehended until 1848). This was a significant factor in increasing their popularity, and exotic ferns began to be grown in botanic gardens and in the glasshouses of the wealthy throughout the country.

Opposite: Detail from a postcard of 'Carrying Fern' from Robert Mclan's *Highlanders At Home* (1848). Selling bracken ashes for soap and glass manufacture provided an income for crofting families in Scotland. Bracken was also collected for use as animal litter or thatch.

The *Gardeners' Chronicle* was the major gardening journal of the Victorian period. Founded in 1841 by Joseph Paxton and John Lindley, it often featured articles on ferns and ferneries.

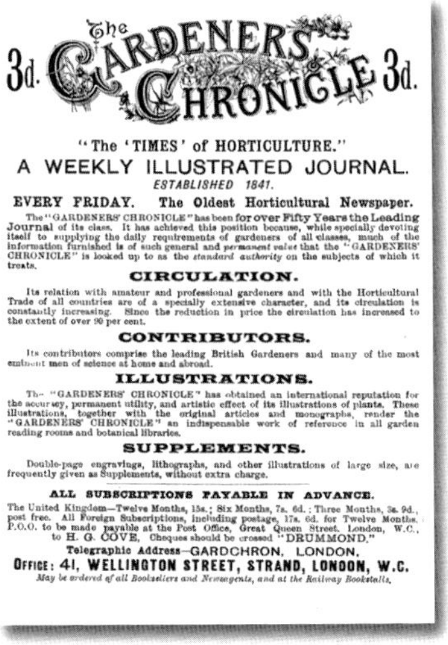

3d. The GARDENERS' CHRONICLE 3d.

"The 'TIMES' of HORTICULTURE."

A WEEKLY ILLUSTRATED JOURNAL.

ESTABLISHED 1841.

EVERY FRIDAY. The Oldest Horticultural Newspaper.

The "GARDENERS' CHRONICLE" has been for over Fifty Years the Leading Journal of its class. It has achieved this position because, while specially devoting itself to supplying the daily requirements of gardeners of all classes, much of the information furnished is of such general and *permanent value* that the "GARDENERS' CHRONICLE" is looked up to as the *standard authority* on the subjects of which it treats.

CIRCULATION.

Its relation with amateur and professional gardeners and with the Horticultural Trade of all countries are of a specially extensive character, and its circulation is constantly increasing. Since the reduction in price the circulation has increased to the extent of over 90 per cent.

CONTRIBUTORS.

Its contributors comprise the leading British Gardeners and many of the most eminent men of science at home and abroad.

ILLUSTRATIONS.

The "GARDENERS' CHRONICLE" has obtained an international reputation for the accuracy, permanent utility, and artistic effect of its illustrations of plants. These illustrations, together with the original articles and monographs, render the "GARDENERS' CHRONICLE" an indispensable work of reference in all garden reading rooms and botanical libraries.

SUPPLEMENTS.

Double-page engravings, lithographs, and other illustrations of large size, are frequently given as Supplements, without extra charge.

ALL SUBSCRIPTIONS PAYABLE IN ADVANCE.

The United Kingdom—Twelve Months, 15s.; Six Months, 7s. 6d.; Three Months, 3s. 9d., post free. All Foreign Subscriptions, including postage, 17s. 6d. for Twelve Months. P.O.O. to be made payable at the Post Office, Great Queen Street, London, W.C., to H. G. COVE, Cheques should be crossed "DRUMMOND."

Telegraphic Address—GARDCHRON, LONDON.

OFFICE: 41, WELLINGTON STREET, STRAND, LONDON, W.C.

May be ordered of all Booksellers and Newsagents, and at the Railway Bookstalls.

Crucially, however, the nineteenth century saw a fundamental change in the practice of gardening. It became not just the pastime of the rich, but also the passion of the new middle classes. The rapidly increasing urban population resulted in the birth of 'amateur gardeners', who wanted to implement all the gardening crazes at their houses or villas in the towns and suburbs. Even the working classes often had a small front garden or back yard to 'beautify'. They were assisted by technical and chemical advances, together with endless advice and instruction in a proliferation of new gardening literature.

The fernery in the New Zealand Court at The Colonial and Indian Exhibition, London, 1886, from *The Illustrated London News.*

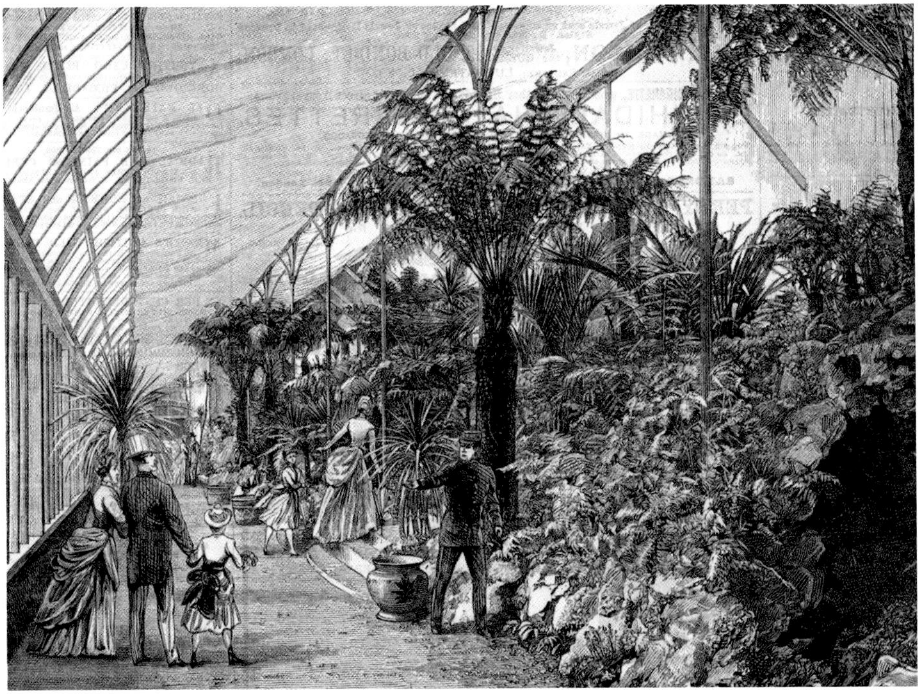

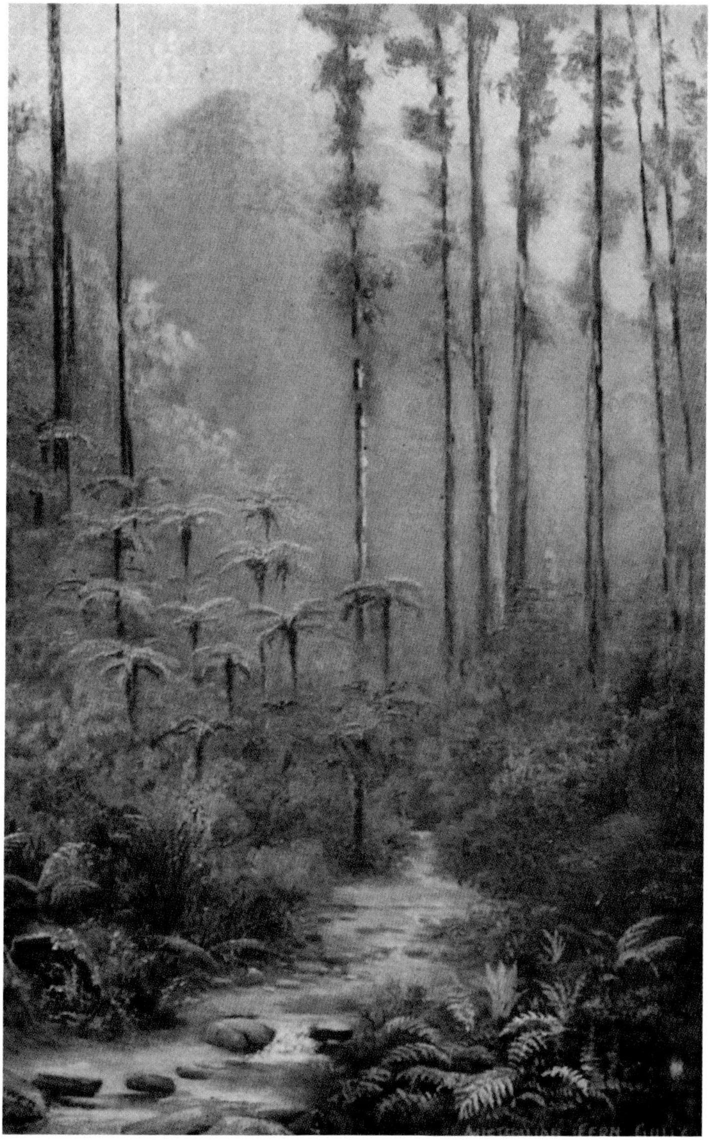

Romantic Australian fern gullies, as depicted in this early twentieth-century postcard, appealed greatly to immigrants and visitors to the country. Eugene von Guérard's *Fern Tree Gully in the Dandenong Ranges,* exhibited at the 1862 London International Exhibition, excited huge interest.

The nineteenth century was also the age of the plant collector, when thousands of new species were introduced to Britain. In part this was led by scientific curiosity, but commercial nurseries also wanted new plants to sell to the ever-increasing numbers of gardeners. Their love of exotics was greatly inflamed by the growth of the British Empire. Illustrations of its fauna and

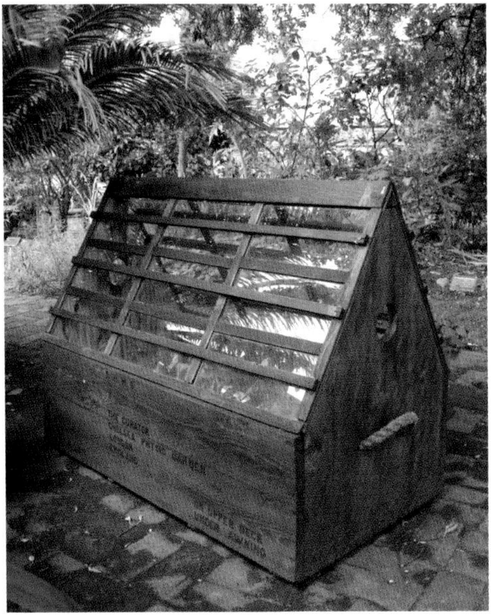

A reproduction travelling Wardian case in Chelsea Physic Garden. One of the most important outcomes of Ward's discovery was the successful transportation of plants around the world. Hundreds of new species, including ferns, were imported to Britain in such cases.

flora were published in books and periodicals, and displayed at exhibitions. Colonialists living and working in countries such as Australia, New Zealand, South Africa and India wanted to recreate what they had seen there when they returned 'home'. In addition, colonial botanic gardens were run by British (often Kew-trained) curators, some of whom took their knowledge of ferns out to places like Ceylon (now Sri Lanka), Mauritius or Jamaica, wrote books on the ferns of those countries, and sent plants back to Britain.

This was made a vastly more successful endeavour by the invention of the Wardian case. Nathaniel Bagshaw Ward (1791–1868) was a doctor who inherited his father's practice in Wellclose Square in the east end of London. He was a keen amateur naturalist, and very early admirer of native ferns. Like other city gardeners, he had to contend with terrible pollution that killed off his plants. He had almost despaired when, in 1829, an incident occurred that had profound results. Ward had put the chrysalis of a Sphinx moth in a sealed glass bottle, and one day noticed that a seedling fern had appeared on the surface of the moist mould in it. It had germinated successfully because it was in an atmosphere free of soot and it had light, heat and moisture; in the evenings the evaporation from the plant condensed on the glass and dropped down like dew.

In 1831 Ward began experiments on a more ambitious scale with the help of the nurseryman George Loddiges (1786–1846), and two years later they had successfully grown thirty species of fern. Although others had been experimenting with raising plants in 'closely glazed cases', it was Ward who realised their potential and did most to publicise the principle. The Wardian case became one of the most important elements in the development of the fern craze.

In 1837 George William Francis published *An Analysis of the British Ferns and Their Allies*, the first popular book on the subject in the nineteenth century. Francis (1800–65) served a 'gentleman's apprenticeship' at Loddiges's nursery, and went on to become the first director of the Adelaide Botanic Garden in Australia. Although it only had plain boards, the book was illustrated with fine black-and-white copperplate engravings by the author. Significantly, it contained a brief mention of Ward's experiments, which was expanded upon in later editions.

However, it was Edward Newman who really started the flood of popular fern literature. Newman (1801–76) dedicated his life to natural history and among his many books wrote *A History of British Ferns* (1840). In this he lavishly praised the work of his neighbour Dr Ward, and wrote that 'The cultivation of Ferns is becoming a fashionable pursuit … almost every one possessing good taste has made … an attempt to rear this tribe of plants.' Fern fever had begun. In 1844 Newman produced a much-enlarged and revised version of his book, *A History of British Ferns, and Allied Plants*, and various editions were published until 1884. Up until the First World War hundreds of books on collecting and cultivating ferns (and what were known as the fern allies: club-mosses (*Lycopodiums* and *Selaginellas*), pepperworts (*Marsileaceae*) and horsetails (*Equisetums*)) appeared, many with very attractive pictorial covers and fine illustrations.

The fern craze affected all classes of society, and had followers among men, women and children throughout the British Isles as well as, to a lesser degree, America and the British Empire. It was at its height between 1850 and 1890, and in 1855 the Reverend Charles Kingsley appears to have coined the name by which this love of pteridophytes became known. In his book *Glaucus; Or, the Wonders of the Shore* he wrote:

> Your daughters, perhaps, have been seized with the prevailing 'Pteridomania,' and are collecting and buying ferns, with Ward's cases wherein to keep them, (for which you have to pay), and wrangling over unpronounceable names of species, (which seem to be different in each new Fern-book that they buy,) till the Pteridomania seems to you somewhat of a bore.

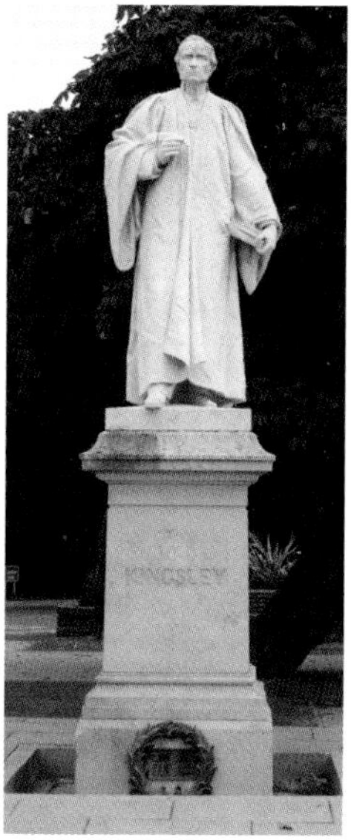

Left: Statue of Charles Kingsley (1819–75), who coined the term 'Pteridomania', in Bideford, Devon.

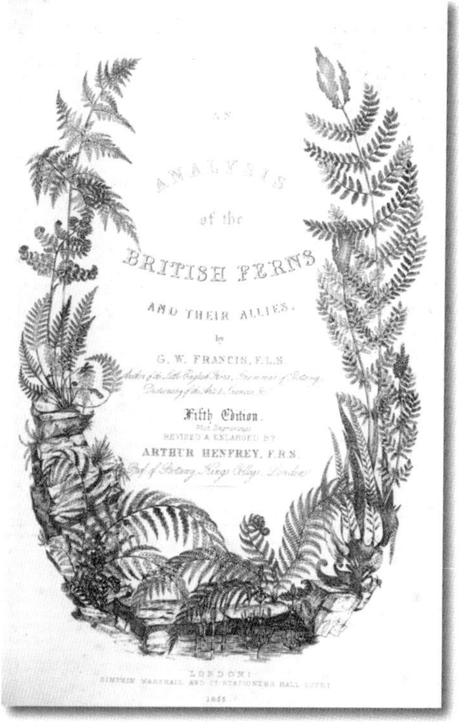

Below: The title page of the fifth edition of G.W. Francis's *An Analysis of the British Ferns and Their Allies*. A thousand copies of the first edition were sold in two years.

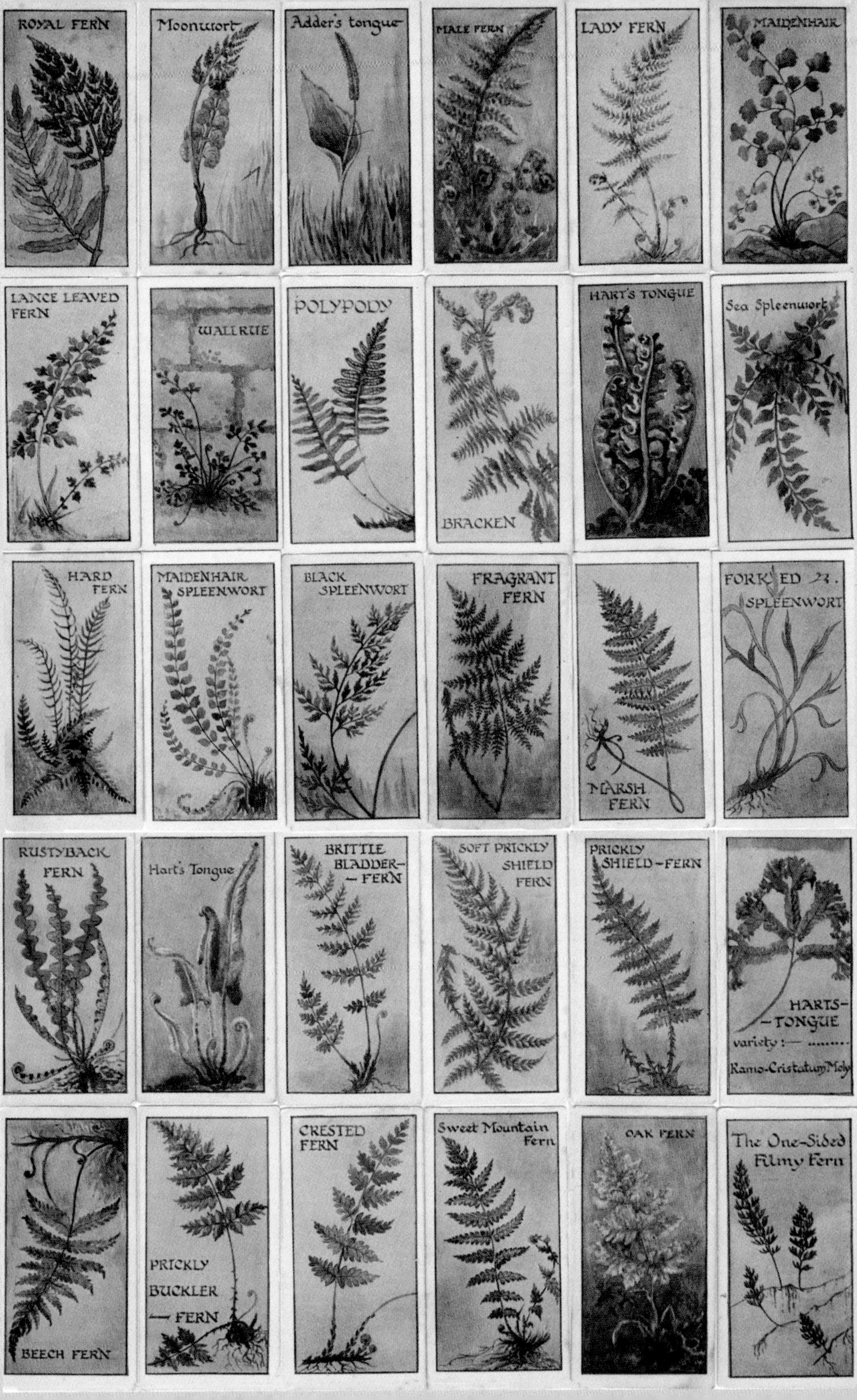
ROYAL FERN
Moonwort
Adder's tongue
MALE FERN
LADY FERN
MAIDENHAIR
LANCE LEAVED FERN
WALLRUE
POLYPODY
BRACKEN
HART'S TONGUE
Sea Spleenwort
HARD FERN
MAIDENHAIR SPLEENWORT
BLACK SPLEENWORT
FRAGRANT FERN
MARSH FERN
FORKED
SPLEENWORT
RUSTYBACK FERN
Hart's Tongue
BRITTLE BLADDER-FERN
SOFT PRICKLY SHIELD FERN
PRICKLY SHIELD-FERN
HARTS-TONGUE
variety :— ....
Ramo-Cristatum Moly
BEECH FERN
PRICKLY BUCKLER FERN
CRESTED FERN
Sweet Mountain Fern
OAK FERN
The One-Sided Filmy Fern

COLLECTING FERNS

THE FIRST WAVE of enthusiasm for native ferns came as much from amateur field botanists as gardeners. Natural history was a national obsession among the upper and middle classes of Victorian Britain, fundamentally because it was believed that examining the 'Wonders of Creation' would 'Lead through Nature up to Nature's God.' It was constantly avowed that there was no better group of plants to study than ferns in order to accomplish this. Numerous fern books were published by the Society for Promoting Christian Knowledge (SPCK) or written by clergymen. Indeed, the nineteenth century was the heyday of the 'parson naturalist'.

Of all the natural sciences, botany was considered the most suitable for middle-class ladies to study. Many fern books were written or illustrated by them, particularly those aimed at other women, beginners, the 'general reader', or children. This was especially so in America, where authors included Margaret Slosson, Mabel Osgood Wright and Helen Eastman. In England Anne Pratt wrote and illustrated numerous works, including *The Ferns of Great Britain and Their Allies*, and Ada Brooke illustrated Francis George Heath's *British Fern Varieties*.

But the authority on native ferns was Thomas Moore (1821–87). His first book, *A Handbook of British Ferns* of 1848, was a very small volume with simple woodcuts, but it was considered the standard work for a long time, with enlarged editions in 1853 and 1857. In 1851 Moore produced *A Popular History of the British Ferns and the Allied Plants*. It was one of a series of cheap, but illustrated, popular handbooks on natural history intended for beginners, and contained twenty hand-coloured plates by W. H. Fitch, one of the foremost botanical illustrators in Britain.

This work has been described as a minor classic. There were second and third editions and then, from 1859 to 1867, seemingly endless editions of an abridgement of it in three different formats. These were called *British Ferns and Their Allies: An Abridgment of The 'Popular History of British Ferns,' And Comprising the Ferns, Club-mosses, Pepperworts, and Horsetails* (in superior or cheap editions) or just *British Ferns and Their Allies: Comprising the Ferns, Club-mosses,*

Opposite: Devon was so inextricably linked with ferns that as late as 1927 James Pascall, chocolate and confectionery manufacturers, issued a set of thirty trade cards featuring 'Devon Ferns'.

Right: 'The cheerful community of the polypody', from Frances Theodora Parsons's *How to Know the Ferns* (New York, 1899).

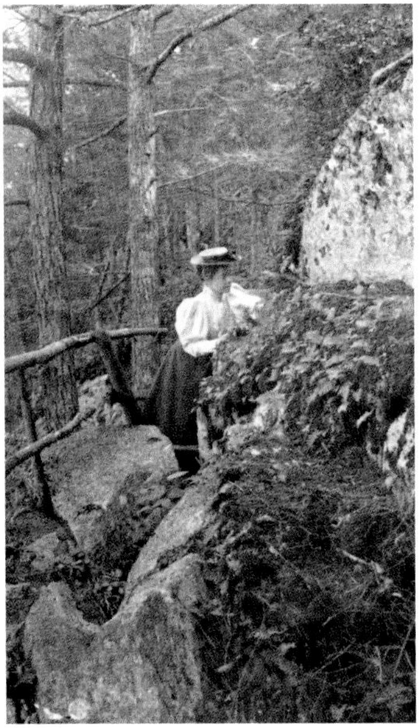

Far right: Third edition of Thomas Moore's *A Popular History of the British Ferns and the Allied Plants, Comprising the Club-mosses, Pepperworts, and Horsetails* (1859).

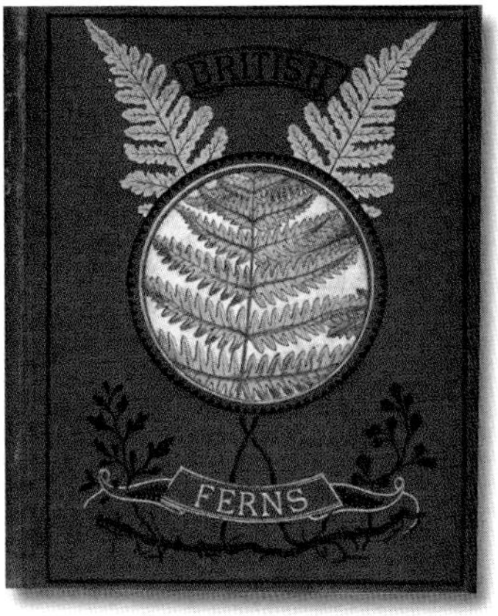

Pepperworts, and Horsetails. Even though the cheap version of the 'abridgement' cost one shilling compared to three shillings and sixpence for the superior edition, the text was exactly the same. And although the twelve plates were initially black and white, these were soon replaced with colour. All in all, it was a bargain and being pocket-sized was easily carried in the field.

At the other end of the spectrum was the elephant folio-sized edition of Moore's *Ferns of Great Britain and Ireland, Nature-printed* (1855–6), priced at six guineas. Although the text was up to Moore's usual standard, the book's

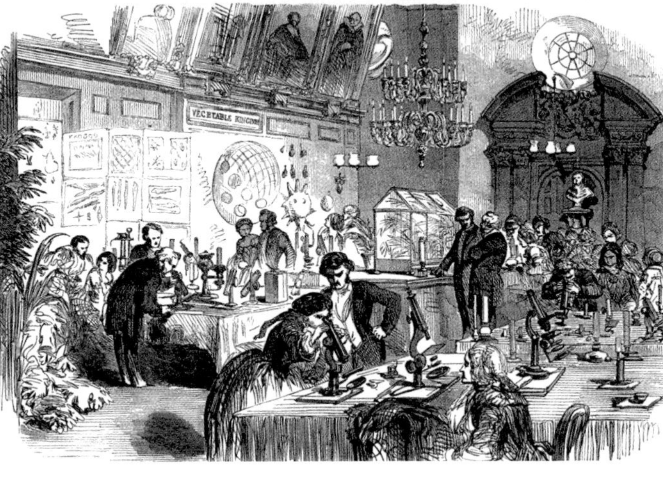

A 'Scientific Conversazione' given by Dr Ward at Apothecaries Hall in 1855: a Wardian case takes centre stage. Microscopes enabled the visitors 'to behold the greatness and wisdom of the Almighty, as displayed in the minutest objects of creation.' (*The Illustrated London News*).

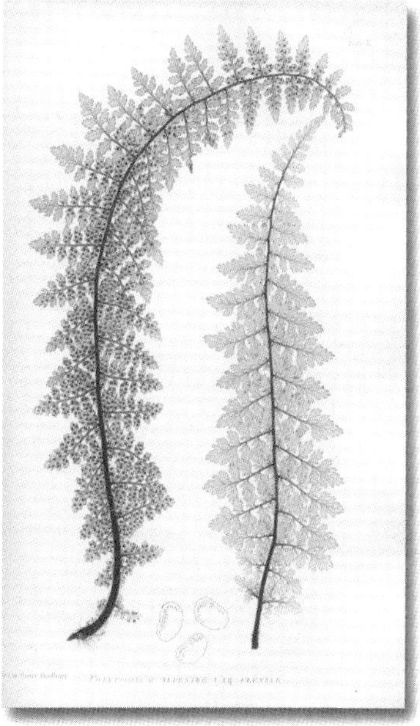

Nature-prints were created by making an impression of a dry plant in soft lead, which was then used as a printing plate, transferring the image by electrotype to a copper plate, which was hand-inked in colour and printed in intaglio.

importance came from the fifty-one beautiful plates by Henry Bradbury (1831–60), who had patented a method of nature-printing. This delineated the structure of ferns very clearly and accurately as it reproduced the finest details. Moore's *Octavo Nature-Printed British Ferns* (Part I 1859, Part II 1860) contained 113 plates by Bradbury who by then had improved the technique.

The public had a rapacious appetite for fern books and titles poured out. Often, as with Nona Bellairs's *Hardy Ferns: How I Collected and Cultivated Them* (1865), they were written in a moralising, sentimental and (ironically) flowery style, with much local 'colour' thrown into descriptions of fern hunting. Of course their quality also varied widely and some efforts, such as Margaret Plues's *Rambles in Search of Ferns* (1861), and *A Plain and Easy Account of the British Ferns* (1854), by the Reverend Edwin Bosanquet, received very bad reviews.

All these books, together with articles in periodicals, urged everyone to go fern hunting, whether for a few hours or weeks. This activity was depicted as not only pleasurable but also moral, healthy and educational. However, there was a less-alluded-to benefit, as *Punch* observed in 1869: 'Botanising is not a bad way of getting over the afternoon, and if you can get your basket well fern-ished so much the better; and it is a well-known fact that the rarest specimens grow in the least frequented spots, so you and your blooming companion can – but the hint is sufficient.'

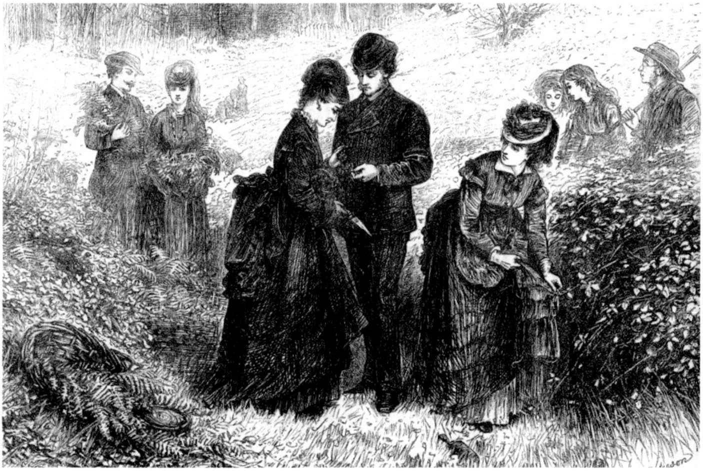

'Gathering ferns' by H. Paterson, from *The Illustrated London News*, 1 July 1871. Fern collecting was a good excuse for mixing with members of the opposite sex.

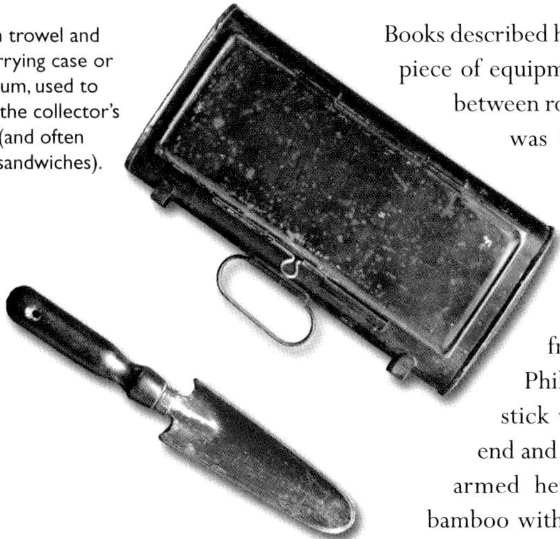

A fern trowel and tin carrying case or vasculum, used to carry the collector's ferns (and often their sandwiches).

Books described how to prepare for a foray. An essential piece of equipment was a narrow fern trowel to dig between rocks and extract a find with its roots. It was then consigned to a vasculum, the botanist's tin collecting case (although some people used a basket, carpet bag or 'tourist's bag'). The receptacle would be lined with wet cloth or moss to keep the plants fresh. The fern collector William Phillips's requisites included a walking stick with a 6-inch-long steel pick on one end and a hammer on the other. Nona Bellairs armed herself with a 15-foot-long piece of bamboo with a knife tied on the end to reach the rare Sea spleenwort (*Asplenium marinum*) high up in a Pembrokeshire cave in Wales. She also recommended taking a book of real fronds or 'patterns' to which to match finds, and a magnifying lens to study the sori (the clusters of sporangia that contain the spores) as an aid to identification.

The passionate pteridologist Charles Druery (1843–1917) found that a vasculum got in the way while he was clambering about, so took 'a ball of string, a few newspapers, a stout reliable hooked stick, and the inevitable trowel'. These instruments were not enough, however, when he was faced with a 1½ cwt mass of *Lastrea montana*. He had to employ a labourer with a fork and spade to dig it up, and a man in a cart to convey it on the first stage of its journey from Devon. Once, a fern was growing so deeply in a rock chink that 'the finder had to roll the rock itself across country and actually grow the Fern out of its hermitage at home.'

'"Splendid ferns everywhere," says Miss Netley, who is seldom out without an Alpine-stick and a basket, as if she were going to market.' ('A Devonian Period', a weekly account of a stay in Ilfracombe, *Punch*, 14 September 1889).

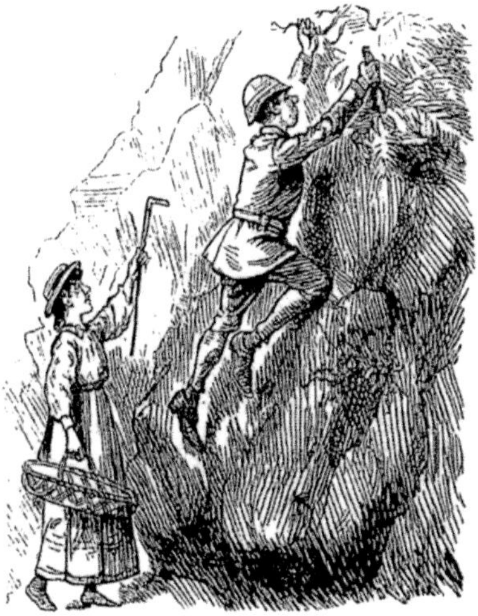

On another occasion, he dislodged a 'very desirable' hart's tongue fern from an arch spanning a Devonshire stream by using a trowel lashed to a stick, and caught it in an open umbrella suspended under the arch by a string. Another hart's tongue 10 feet up a wall could only be secured by the use of a village lad hoisted upon Druery's shoulders. This overwhelming desire to 'capture' a rare fern led the hunter or huntress to wade through streams, scale rock faces, descend gorges and lean over fast-flowing rivers. Accidents were common, and sometimes even fatal.

There are only about fifty species of fern native to Britain (of nearly 13,000 in the world altogether), and in 1837 Francis had claimed to illustrate every species and variety. However, by the mid-nineteenth century it was realised that British ferns produce hundreds of varieties. Collectors started hunting for these 'sports', a number of which were so bizarre, with multiple forked and crisped fronds, that their detractors called them 'monstrosities'. In *Our Native Ferns: Or A History of the British Species and Their Varieties* (vol I 1865, vol II 1867), the fern breeder Edward Lowe described 1,294 varieties.

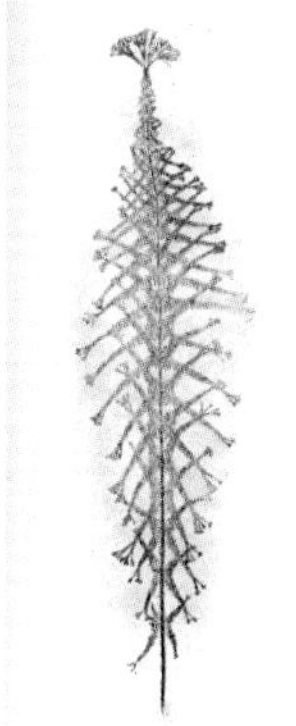

The amazing cruciate and crested variety of lady fern called 'Victoriae' after Queen Victoria. Further variations developed by Edwin Fox were called *Victoriae magnificum* and *Victoriae gracile*.

The most prolific producer of varieties is the lady fern (*Athyrium filix-femina*), and a stunning cruciate form was found near Loch Lomond in Scotland in 1861 and named 'Victoriae' in honour of the Queen. The hart's tongue (*Asplenium scolopendrium*) is also extremely prone to sport. Some varieties are still popular and widely available: *Dryopteris affinis cristata*, 'The King', was found in Cornwall some time before 1850 and is still sold in garden centres today. The leaders of the hunt for varieties named each 'new' fern found, even though it could be a unique example! They usually added a third, descriptive, Latin name after the genus and species such as *digitatum*, *laceratum*, *deficiens*, *undulato*, *irregulare* or *cristatum*. However, as ever more varieties were found compound descriptions such as *congesto-cristatum* were introduced, synonyms used, Latin terms invented and classical names such as 'Athene' appended. Some varieties were named after the person who found them, such as *Athyrium filix-femina* 'Frizelliae', discovered by Mrs Charles Frizell in 1857.

In 1890 Lowe published *British Ferns and Where Found.* In very small type the book listed every variety known at that time – all 1,861 of them – who had found it, and where. The descriptions were of necessity so brief that it cannot have been useful, and in fact must have been rather off-putting, to the 'young collector' at whom it was aimed. However, in his book Lowe used a system of nomenclature originally devised by Dr F. W. Stansfield, which remained unaltered until the 1960s. This discouraged the adoption of compound Latin names, although the practice continued.

Fronds were often gathered for a herbarium. Specimens were pressed between sheets of botanical drying-paper, blotting-paper, or old newspapers, sandwiched between two boards with leather straps drawn tightly around them. Then they were sewn, glued or secured by thin, gummed strips of paper on to leaves in an album. The fronds were labelled with their names, where found, and the date. This was often recommended not just for botanical reasons but also for the memories of the 'happy hours spent with loved and valued friends' that such a record would invoke.

Publishers soon latched on to this market. An advertisement in the front of *British Ferns with Illustrations* (1860) by Mrs Phebe Lankester, announced 'THE FERN COLLECTOR'S ALBUM: a descriptive folio for the reception of Natural Specimens, containing on the right-hand page a description of each Fern, printed in colours; the opposite page being left blank for the collector to affix the dried specimen, forming when filled an elegant and complete collection of this interesting family of plants.'

Printed labels were also included in books. As early as 1845 Newman published *A Catalogue of British Ferns . . . Intended for Labels,* probably in the vain hope of introducing greater uniformity to the nomenclature of ferns. Marian Ridley produced *A Pocket Guide to British Ferns* (1881), which included 'a Set of Labels for the Herbarium' at the end. Sadie Price's *The Fern-Collector's Handbook and Herbarium* (1897), contained no text, just black-and-white drawings of ferns of the northern United States. All the pages were printed on one side only and the book was bound with double card strips in the folds of alternate plates to allow for ferns being added in.

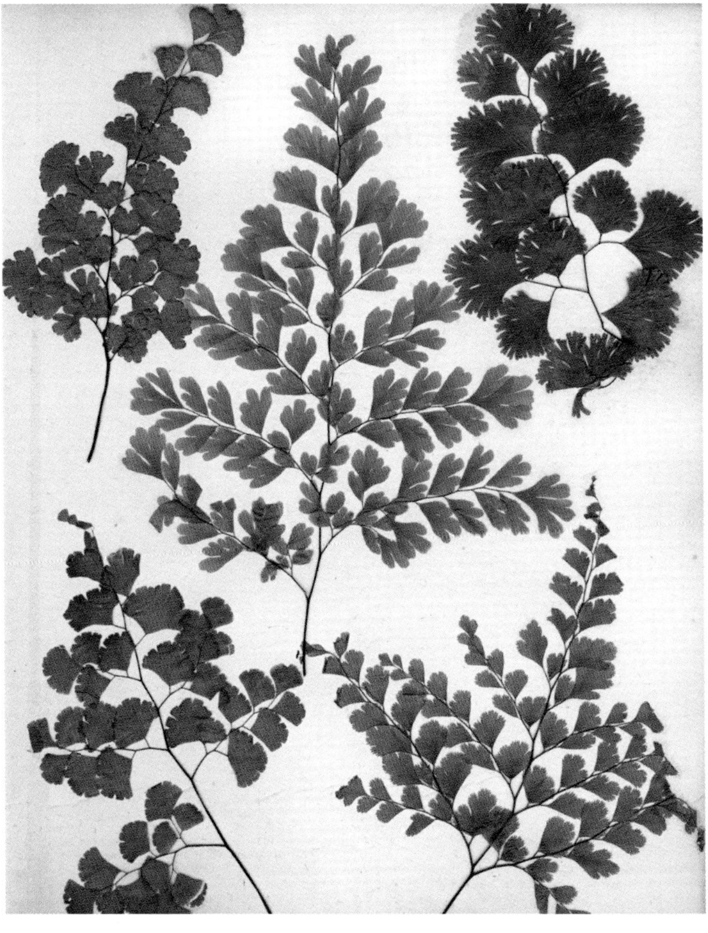

A page of pressed varieties of maidenhair ferns from an 1870s fern album.

Some albums, such as Jane Patison's *Gleanings Among the British Ferns; Illustrated by Dried Specimens* (1858), even came ready-made with sets of labelled, pressed plants. Not surprisingly, the areas covered by these albums were generally those best known for ferns, such as *Ferns of Wales*, *The Ferns of Moffat*, and *Flintoft's Collection of the British Ferns in the English Lake District*. They were offered as a tasteful, beautiful and scientific souvenir to those who had either been unsuccessful in their searches, or did not care to trawl the countryside themselves.

Ferns grow most prolifically in the warmer and wetter parts of Britain, and those regions (to which the ever-growing railway network provided improved access for the masses), became hot spots of fern activity. They attracted collectors, formed the subject of books and articles, and saw the establishment of fern nurseries. Scotland was one such area, also Derbyshire, Yorkshire, Shropshire and Staffordshire. However, *the* Fern County was Devon. In the 1850s the writings of Charles Kingsley (1819–75) and the religious naturalist Philip Gosse (1810–88) helped establish the north Devon coastal resorts of Ilfracombe, Lynton and Lynmouth as popular holiday destinations. Visitors spent their time profitably, investigating rock pools or collecting ferns. Kingsley's sister Charlotte was the wife of well-known pteridologist the Reverend John Mills Chanter of Holy Trinity Church, Ilfracombe, and in 1856 she attracted more visitors with *Ferny Combes: A Ramble After Ferns in the Glens and Valleys of Devonshire*.

South Devon was not ignored; in 1862 the Reverend Zachary Edwards published *The Ferns of the Axe and its Tributaries*, and the eccentric antiquary Peter Orlando Hutchinson produced *The Ferns of Sidmouth Illustrated With*

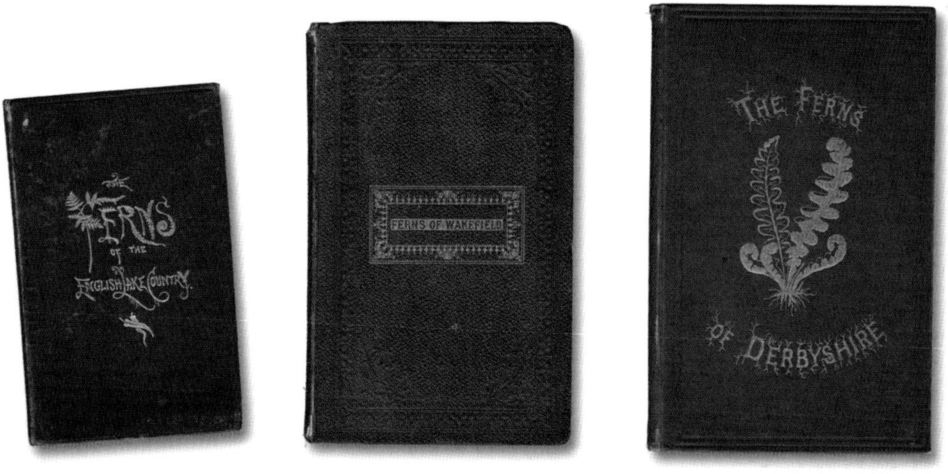

Many books on the ferns of particular areas were written to guide the collector.

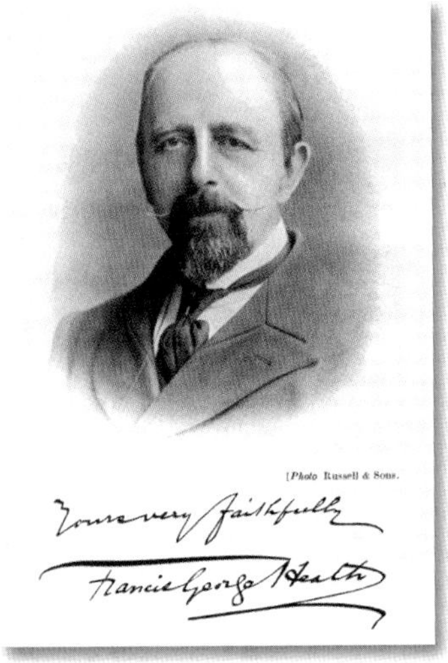

The fern book author Francis George Heath (1843–1913).

Impressions of the Ferns Themselves. The author who did most to advertise Devon's charms was the populist writer Francis George Heath whose books were very much for the amateur fern lover. His first effort, *The Fern Paradise: A Plea for the Culture of Ferns* (1875), ran to eight editions (the last one in 1908). *The Fern World* (1877), ran to twelve editions, the last published in 1910. Heath described the charms of rambles through Devonshire, or 'Fernland', conjuring up irresistible images that encouraged many to follow in his footsteps.

The engraver William Linton, author of *The Ferns of the English Lake Country* (1865), claimed that *this* district might be called 'the Land of Ferns'. In fact his publisher, John Garnett of Windermere (who also offered an album of 'Ferns From the Lakes'), was one of ten fern enthusiasts who gathered in Kendal in September 1891 to found the Northern British Pteridological Society. There had been a few previous attempts at founding such societies: in about 1858 William Phillips started the Wardian Society in Belfast, but it did not survive long, and another Pteridological Society was founded in 1871 but only lasted until 1876.

Within a year members from the whole of Britain had joined the new society, and at the first AGM 'Northern' was dropped from its name. By 1899 the membership only stood at forty-five and it was decided to produce *The Book of British Ferns* (1903), compiled and edited by the second president

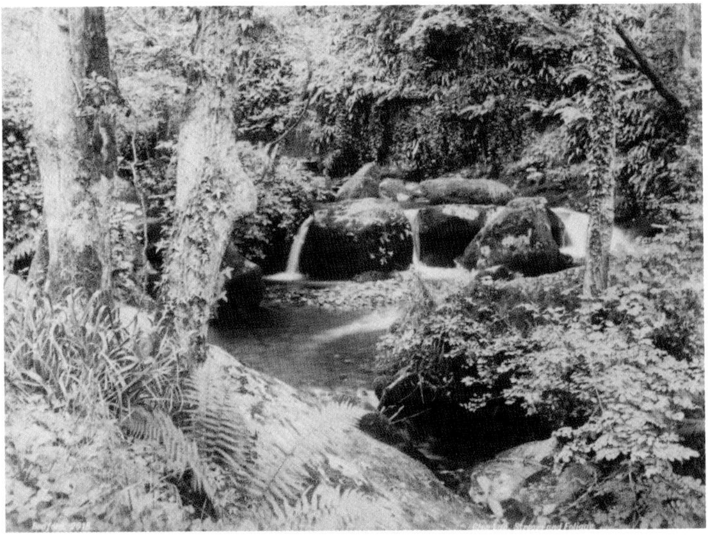

Devon delights: 'We look in wonder upon a fairy, dreamy scene of clustering ferny forms in fascinating association with mossy rocks and flowing water.' (Francis George Heath).

Charles Druery. This may have encouraged new blood, and at the 1909 AGM around 120 new members, including about a dozen women, joined. During this period, however, the society's activities remained modest. The American Fern Society was founded in 1893 and many local groups such as the Leek and Moorlands British Fern Society were also formed. Members entered the numerous horticultural shows in the provinces and London, which had various classes for ferns.

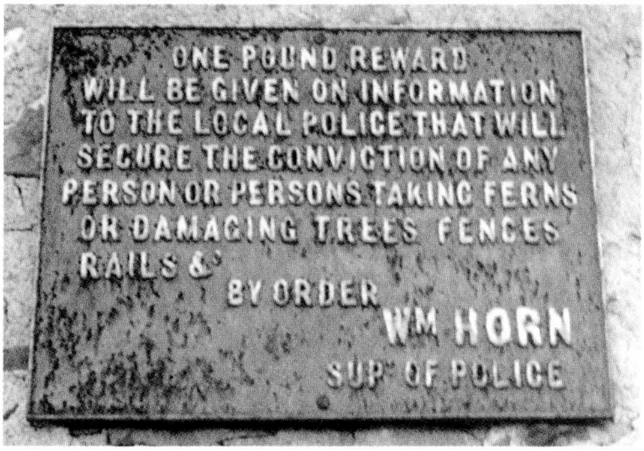

A sign warning against fern stealing, near Ingleton, Yorkshire.

The fern experts – many of them members of the British Pteridological Society – saw themselves as leaders of the fern *cult* rather than *craze*. Some justified their collecting with the plea that it helped to increase knowledge of the distribution of the plant. Others claimed that finding varieties and putting them into cultivation helped preserve them for posterity. They were at pains to distinguish their activities from those of the 'fern destroyer', even if they seemed remarkably similar at times.

As early as 1840 Newman was deploring the fact that *Asplenium lanceolatum* had become 'nearly extinct' in the neighbourhood of Tunbridge Wells in Kent 'owing to the ravages of unprincipled botanists.' Worse were the thoughtless 'trippers' who filled their baskets with ferns only to let them die on the way home. Country people also took advantage of the middle-classes' obsession to earn a little money guiding tourists to fern localities. By all accounts, this was a popular occupation for local children. Alternatively, they kept the source secret and supplied the 'addict' with sought-after varieties, while asserting that it was far too dangerous for the lady or gentleman to find them themselves.

However, while often recommending the services of the commercial fern collector, at the same time authors reserved their greatest censure for 'The Itinerant Fern Vendor' or 'Professional Fern Tout'. These characters scoured the countryside for specimens, sent them up to the cities by rail, and then sold them in the streets or hawked them from door to door. Even in the mid 1890s Lowe despaired that gangs of marauding robbers were plundering the countryside for its ferns. Not surprisingly, Devon was particularly targeted and in 1896 the *Gardeners' Chronicle* reported how, as there was then no bye-law against taking ferns from the wild, two fern stealers had been charged at Totnes with damaging Devonshire hedges. Previous convictions were noted, and as the defendants could not or would not pay the fines, they were sentenced to four and six weeks' hard labour each.

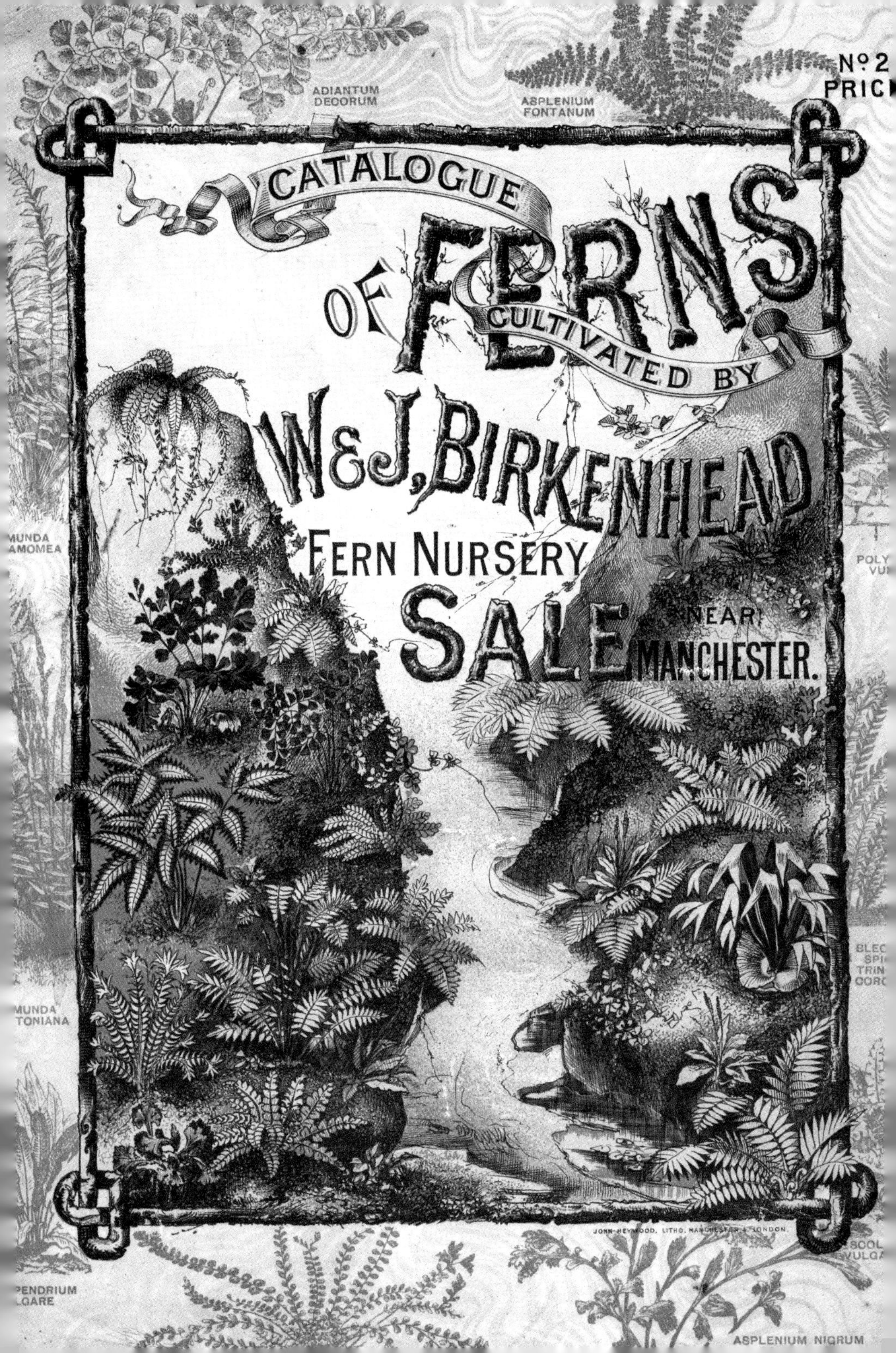

Nº 2
ADIANTUM DECORUM
ASPLENIUM FONTANUM
CATALOGUE
OF FERNS
CULTIVATED BY
W&J, BIRKENHEAD
FERN NURSERY
SALE
NEAR
MANCHESTER.
JOHN HEYWOOD, LITHO. MANCHESTER & LONDON.
ASPLENIUM NIGRUM

CULTIVATING FERNS

WHETHER THE PLANTS had been collected or bought, the fern lover wanted a fernery in which to grow them. This meant one of three things: an indoor glass case, an outdoor area, or a glazed structure. The term was also used to describe the collection itself. Books on collecting ferns often included advice on cultivating them as well. A few fern forays were said not only to infect the hunter with fern fever, but also to teach invaluable lessons as to different ferns' needs. However, many books concentrated on growing ferns, including the nurseryman Benjamin Williams's *Hints on the Cultivation of British and Exotic Ferns and Lycopodiums* (1852), the prolific garden writer [James] Shirley Hibberd's *The Fern Garden* (1869), and John Newsham's *Facts and Fancies About Ferns* (*c.* 1876).

Conrad Loddiges & Sons was the foremost London nursery in the early nineteenth century, and played a leading role in experiments with growing ferns under glass. Ironically, the firm closed in 1852 just as the fern craze really took off. By the 1870s most nurseries sold ferns and many specialised in them. Because the railway could be used to transport plants to customers, some nurseries were in good fern areas, such as north Devon or the Lake District, some distance from major cities. Interest in collecting varieties had led to the practice of propagating them, and by the late 1860s breeders were hybridising ferns by sowing mixtures of spores. More and more cultivars were raised and sold in nurseries with descriptions such as 'distinct and curious'. Their quality varied immensely and there were many inferior specimens on the market.

James Veitch & Sons of Exeter and Chelsea had a cool fernery at their nursery in the King's Road, London, in which they displayed new species brought back by plant collectors. It was overseen by George Schneider, who wrote *The Book of Choice Ferns for The Garden, Conservatory, and Stove* (1890–4). James Backhouse & Son specialised in orchids, alpines and ferns at their 100-acre nursery at Holgate, York. There were forty glasshouses and a rock garden constructed of 1,500 tons of stone. They were particularly known for their filmy ferns (*Hymenophyllums* and *Trichomanes*),

Opposite: Catalogue no. 22 issued by the W. & J. Birkenhead Fern Nursery, Sale, in *c.* 1892. Described by them as 'The largest, most beautiful, useful, interesting and profusely illustrated catalogue of ferns & selaginellas ever published.'

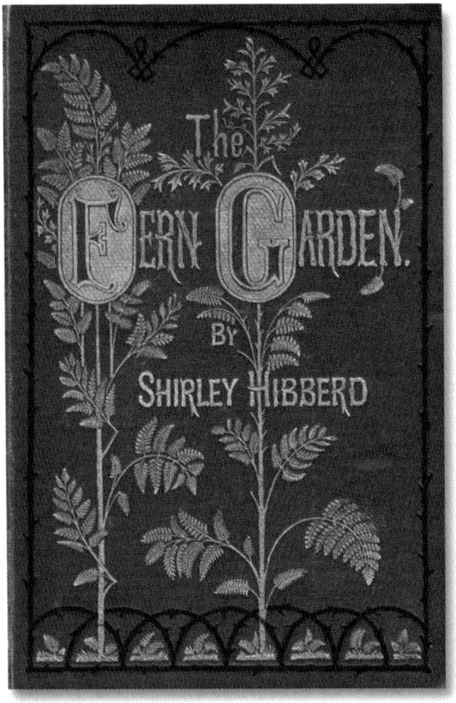

Shirley Hibberd's *The Fern Garden: How to Make, Keep and Enjoy it; or Fern Culture Made Easy*, 1869. The cover depicts very un-fern-like plants.

which they grew in a large, glazed, underground rock garden.

One of the greatest fern nurseries was that of Abraham Stansfield in Todmorden, Lancashire. His first fern catalogue was issued in 1852 and catalogue eight in 1869 offered 1,125 species and varieties of 'Stove, Greenhouse, Hardy Exotic and British Ferns'. William and John Birkenhead's fern nursery at Sale near Manchester operated from *c.* 1868. Their huge catalogues ultimately listed over 2,000 species and varieties. Their fame spread around the world; in 1899 they supplied the American John Morris with ferns for his new fernery in Philadelphia. The brothers also gave talks and published papers on the subject, and in 1892 John Birkenhead wrote *Ferns and Fern Culture*.

Ward's case was publicised during the 1830s, but it was not until 1842 that he fully described his invention in *On The Growth of Plants in Closely Glazed Cases*. This was not illustrated, and its affect on the general public must have been limited. However, in 1845 the onerous excise duty on glass was repealed, prices halved, and glass structures became ever more common. This culminated in the Great Exhibition of 1851, housed in the Crystal Palace in Hyde Park, London. The excitement generated by the Crystal Palace (designed by Joseph Paxton) meant that almost anything made of glass appealed to the populace in the 1850s. In addition, Ward had exhibited cases at the exhibition, designed by his great friend the marine painter Edward Cooke. Cooke was George Loddiges's son-in-law and Loddiges's nursery provided the ferns for the cases. *On The Growth of Plants* was expanded and re-issued in 1852 with delightful vignettes of cases drawn by Cooke and his sister Georgiana, who was Ward's daughter-in-law.

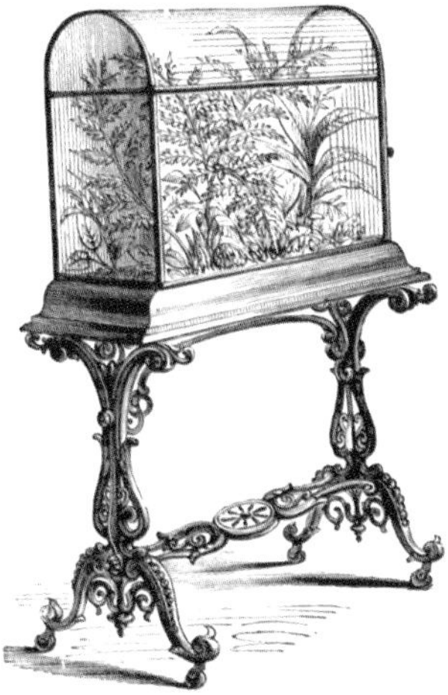

A typical Wardian case on a stand; just over 2 feet long, 1¼ feet wide and 2 feet high. Marketed as both the 'Albert' and the 'Victoria' case.

Wardian cases (initially called Ward's cases) subsequently became ubiquitous in middle-class homes in Britain and America. Because of Ward's own love of ferns, and because they did extremely well in the case's warm, humid atmosphere, ferns were inextricably linked with it. Authors recommended it as *the* way to protect them from the fumes of gas lighting and open fires. As cases were generally placed in the parlour or drawing room, their management was seen as an occupation for ladies, as 'Domestic Floriculture' was generally. They were also recommended for the sick room to cheer the invalid – another area of female responsibility.

The most basic cases comprised a simple, zinc-lined box topped with a glazed frame, which could be made at home. Manufacturers started advertising their services in the early 1840s. The framework for the glass could be made of brass or bronze (which was considered most ornamental), or wood, galvanised iron or zinc, which was cheaper. Books instructed that the top needed to be removable and that there should be a door or two for the owner to be able to give the ferns air and water, for, contrary to popular belief, they did need a little of both occasionally. They also needed drainage, and cases usually had a false base of perforated zinc that was covered with compost and moss. The water collected underneath and was drained via a hole or small tap.

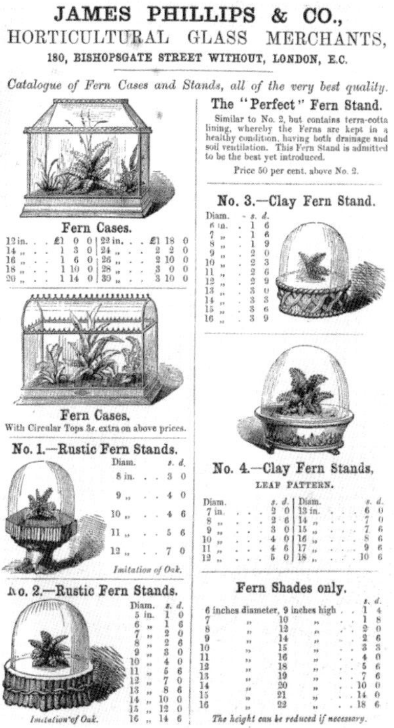

Above: As well as growing ferns in Wardian cases, glass shades or bell glasses over a terra-cotta or earthenware base, and planters with Minton tiles or glass mosaic around them were also popular.

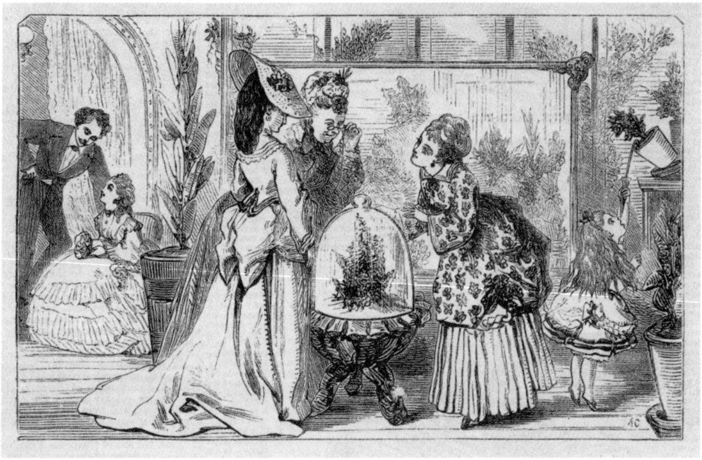

Left: Pteridomania was so widespread that it featured not only in botanic and gardening publications, but also in popular novels and periodicals. This illustration is from *Bow Bells*, 15 November 1871.

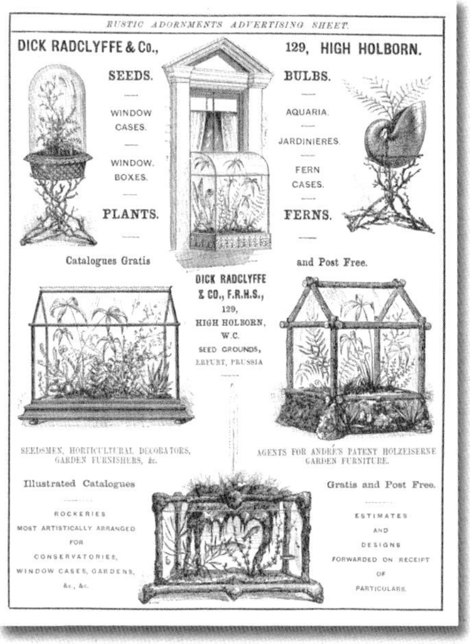

Various homes for ferns advertised by Dick Radclyffe & Co. in Shirley Hibberd's *Rustic Adornments for Homes of Taste* (1856).

Miniature landscapes of 'hills and dells' could be created using coke (coated with Portland cement), virgin cork, or pumice stone. In larger cases these materials could even be used to form arches, walls and 'ruins' covered with planting pockets. This was not only to satisfy a taste for the picturesque or gothic, but also to provide the most extensive range of conditions for the widest variety of ferns. However, writers warned of the dangers of over-filling the case with fossils, coral, shells or minerals, and to remember that 'over-adornment, inside or out, is in bad taste.' Despite such warnings, cases were produced in all shapes, sizes and styles, including Gothic, Rustic, Rococo and Neo-Grec.

Cases could be heated for tender or filmy ferns. Miss Maling's case had a metal cistern in the bottom, into which hot water was poured and let out when cold, 'all the care necessary being to re-fill it with a kettle of hot water night and morning.' The Waltonian case also had a 'boiler' in its base, which was warmed by an oil lamp or gas burner underneath. Others were heated by coils of piping connected to the domestic water-heating system, or sat in pans of hot water.

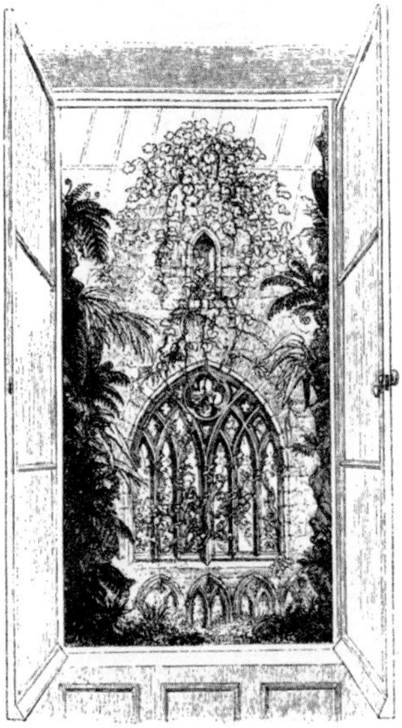

Dr Ward believed that '*tableaux vivans* of the highest interest' could be formed by placing architectural models among ferns in window cases. His 'Tintern-Abbey House' contained a model of the west window of the abbey.

Nathaniel Ward had developed a version of his case, the vivarium, to house snakes and amphibians, and the 'aqua vivarium', or aquarium, for fish. During the 1850s there was a great fashion for parlour aquaria, and they were often combined with Wardian cases. Robert Warington developed 'Robert Warington's Aquatic Plant Case or Parlour Aquarium', which became known as the 'Warrington [*sic*] Case'. It was also often suggested that a Wardian case could be combined with an aviary as the birds and foliage went well together.

Another medical doctor, George Churchill Watson of Liverpool, designed a terracotta planter called the 'Watsonian Patent Fern Brick'. This came in three sizes and had a concave cavity in the middle, in the lower part of which sat a removable planter in the form of a clam shell. The brick was sold by appointed agents F. & G. Rosher. It could be hung up singly, built into walls or a number could be constructed into a 'screen' or a fern column under a glass case known as a Rosher's Pillar.

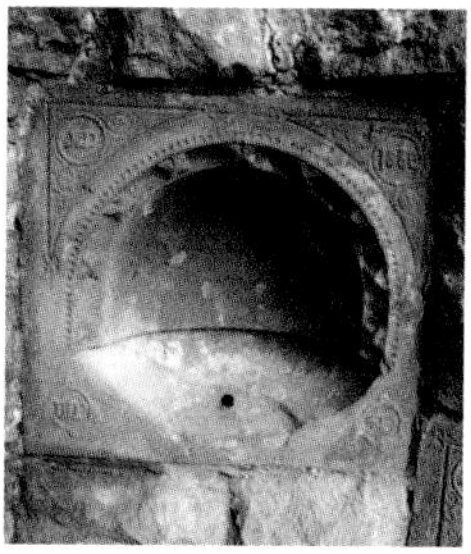

A single Watsonian Patent Fern Brick (missing its shell planter) in the fernery at Newstead Abbey.

Ferns could be grown on landings below skylights, in fireplaces during the summer, in basement 'wells' or 'areas' or on balconies. In fact, all one needed was a window. Window gardening (or *Hortus fenestralis*) originated in Europe and entailed either filling a bay window with plants or attaching a projecting glass case to the exterior of a sash window. It was particularly recommended for the working classes as the vegetation hid bad views or substituted for no view at all.

At the opposite end of the scale were the glass conservatories, winter gardens and hot houses or 'stoves' of the rich. These would house exotics such as orchids and palms, together with tender, foreign ferns, planted among pools, cascades, naturalistic rockwork and the occasional statue. Lord Leverhulme created such a conservatory-fernery at Thornton Manor in Cheshire in the 1890s. The finishing touch was a gallery so as to be able to appreciate the plants from above. Sometimes the fernery occupied only part of the conservatory in the form of a grotto or a rockwork wall constructed especially for the ferns.

On large estates temperate and tropical fern houses could be found forming part of extensive ranges of glasshouses for different plants. As at Cragside in Northumberland, the houses were normally linked so that the visitor progressed through a range of environments. At Gawthorpe Hall in Lancashire an 1870s glasshouse was converted into a grotto-like fernery forming an impressive family and guests' entrance into the glasshouses in the

An Edwardian case, surely one of the largest ever constructed.

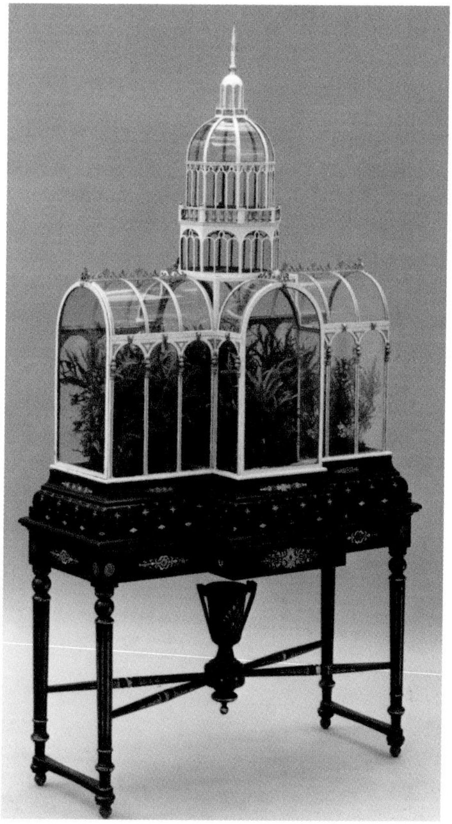

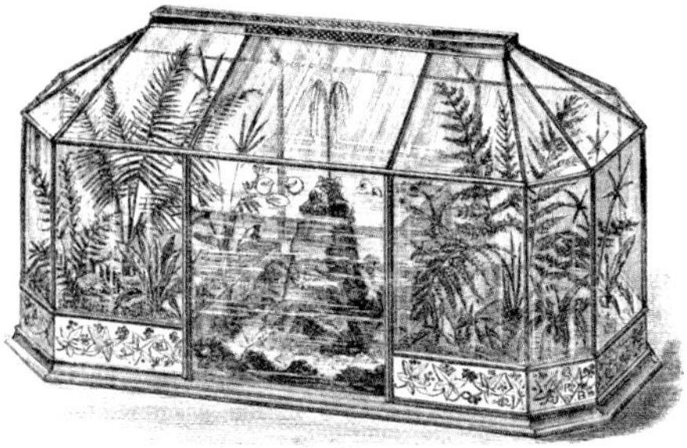

walled kitchen garden. Earlier garden structures were often adapted into ferneries to keep up with prevailing fashions. The 1830s orangery at Gunnersbury Park in Middlesex was made into a fernery housing tree ferns given to the Rothschild family by the governor of Tasmania in 1873.

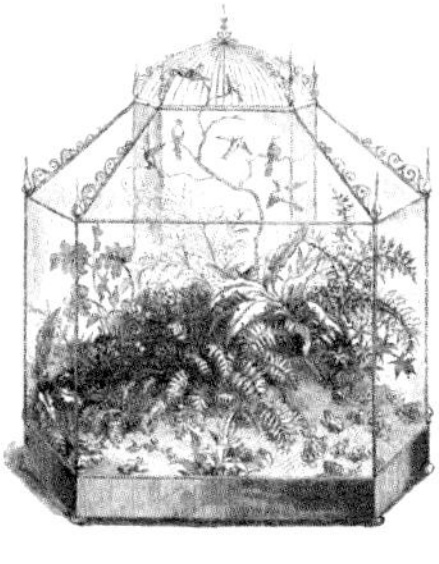

Above: A combined Wardian case and aquarium. As Shirley Hibberd memorably wrote, 'Who would live contentedly, or consider a sitting-room furnished, without either a Ward's Case or an Aquarium?'

Above right: A decorative Wardian case combined with a small aviary, from the *Illustrated Times*, 20 June 1857.

However, the classic fernery was the freestanding, glass-roofed structure usually constructed for members of the upper-middle class in the suburbs and environs of cities like London, Glasgow, Dublin and Belfast. In contrast to conservatory-ferneries, they had high stone or brick walls or were semi-submerged to provide the shade and dampness that native ferns require. If extra shading were needed on the glass, climbers, green-tinted limewash, blinds or muslin could be employed. Occasionally coloured glass was used, particularly green, but this tended to block out too much light. In Australia and New Zealand ferneries had to be shaded with lattice or trelliswork against the intense sun, and were also known as shade houses.

Inside, the principal aim was to recreate a natural fern-clad ravine. Therefore the structure of the building was hidden as much as possible with vegetation and rockwork, and the planting was naturalistic in style. Ferneries could be heated, but hardy native ferns predominated. A fernery was intended to be a cool, shady, oasis of green. The Victorians loved colour – in their clothes and homes as well as gardens – but there were those who

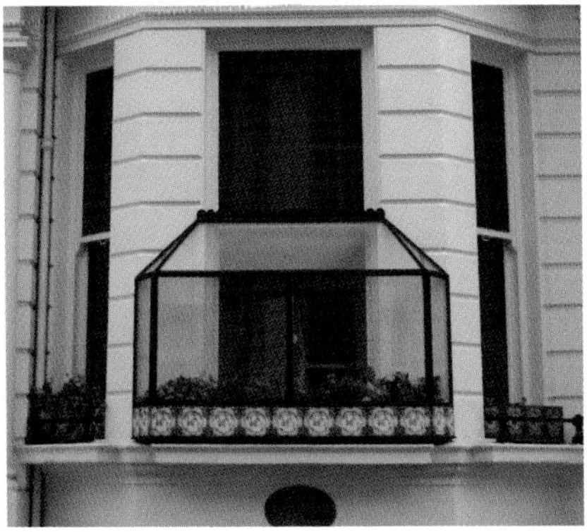

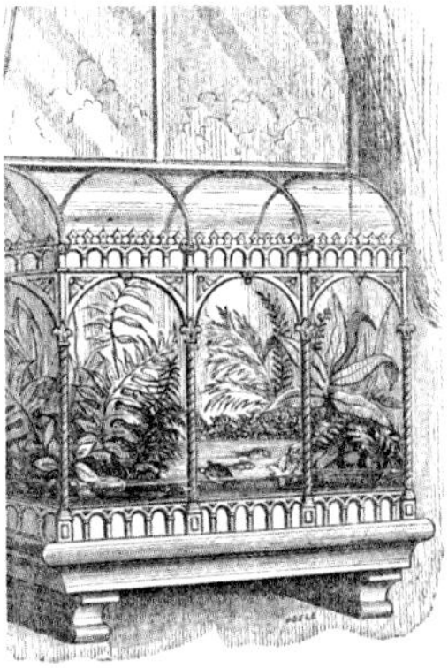

Right: A fern case on the front of Linley Sambourne House, London: a rare surviving example of a once common feature.

Far right: Fern case, aquarium *and* window gardening combined in John R. Mollison's *The New Practical Window Gardener*, 1879.

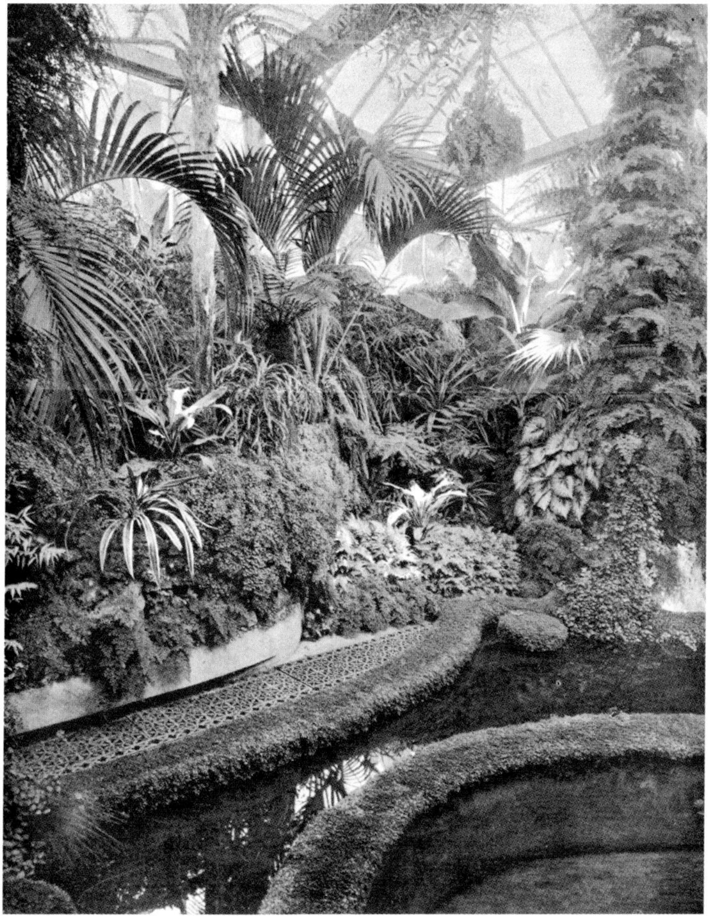

Left: The interior of a typical conservatory-type fernery, with pool and decorative iron grilles covering heating pipes, at Nash Court, Kent.

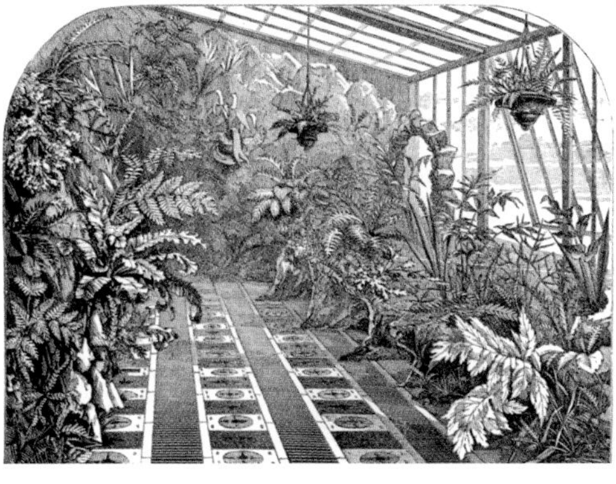

Below: Shirley Hibberd's own fern house at Stoke Newington was typical of the smaller, urban, covered fernery.

decried it as gaudy, and championed foliage rather than garish bedding schemes. Authors and nurserymen proclaimed that a love of the flowerless fern indicated a cultivated intellect and refined taste – and no one wanted to appear vulgar.

Although ferneries could be rectangular, writers recommended an L-shaped or irregular plan to add an element of surprise for the visitor. A meandering path, covered with tiles or

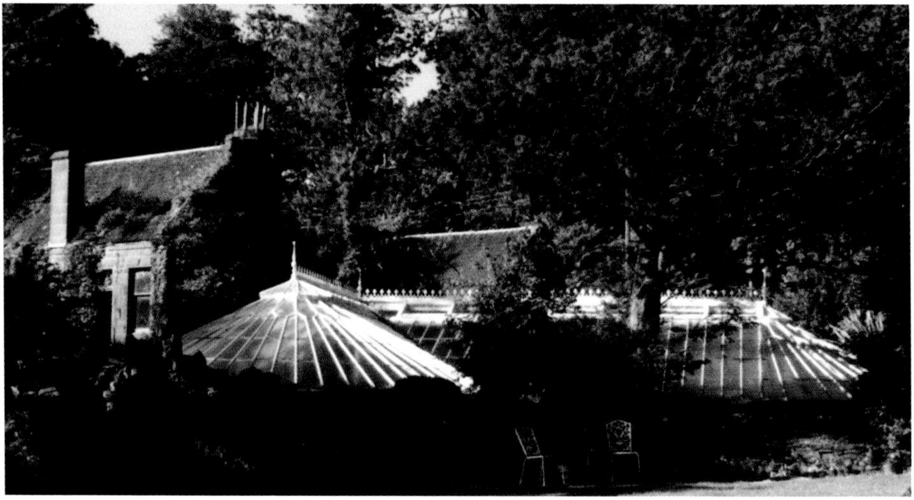

The restored glazed fernery at Ascog Hall on the Scottish Isle of Bute has many of the typical features of this sort of structure. It was constructed for Alexander Bannatyne Stewart in the 1870s. (Photo 1997)

pebble-mosaic, ran around the perimeter between rockwork covering the sides of the building and massed in the middle. In the larger or more elaborate ferneries the stone was built up into arches and 'ruins'. There could be alcoves, caves, grottoes, a subterranean passage, or stairways leading up to viewing platforms. Water was essential, and was usually introduced in the form of a cascade down an end wall. It then flowed through the house in a series of pools crossed by a rustic bridge. The sound of water was important, and this was sometimes provided by way of a 'dropping' or 'dripping' well. As with Wardian cases, these features were introduced to provide a wide

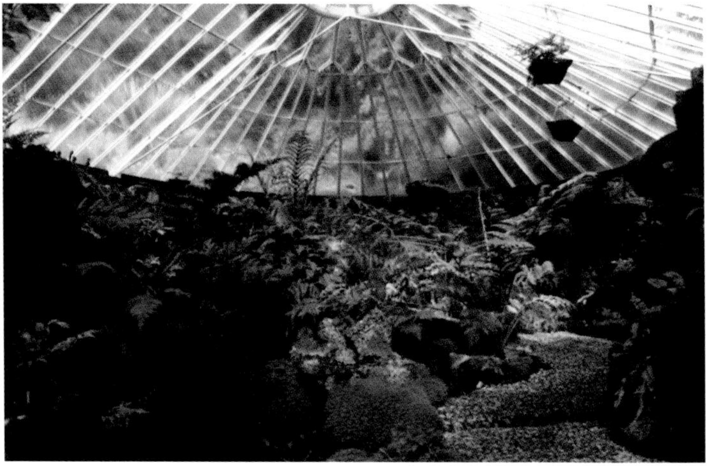

Because it was semi-submerged, Ascog fernery was warm enough for the planting to be subtropical. Tree ferns included *Dicksonia antarctica*; the silver tree fern, *Cyathea dealbata*; *Dicksonia squarrosa*; and the fine, black-stemmed New Zealand tree fern, *Cyathea medullaris*. (Photo 1997)

variety of growing conditions as well as for aesthetic reasons. Ferns could also be planted in hanging wire or twiggy baskets lined with moss. An optional extra was wildlife of some sort: birds, monkeys, goldfish, tortoises, lizards, toads and frogs have all been found in the fernery. Henry Boyle kept alligators in his in the Lake District.

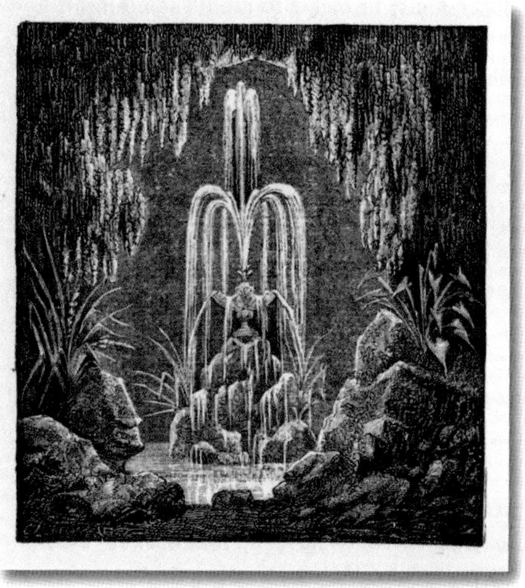

Many Victorians were obsessed with the fairy world, and ferneries are often described as 'fairy grottoes' or a 'fairy land'. In accounts of fern collecting beneficent fairies lead the hunter to wonderful discoveries, while dark fairies present finds at inopportune moments or hide them altogether.

The pre-eminent fernery constructors from the 1860s onwards were James Pulham & Son of Broxbourne in Hertfordshire. Their promotional booklet *Picturesque Ferneries and Rock-garden Scenery, in Waterfalls, Rocky-streams, Cascades, Dropping Wells, Heatheries, Caves or Cavernous Recesses for Boathouses &c., &c.*, published *c.* 1877, lists nearly two hundred works. The Pulhams made use of local stone whenever possible, but if it was not available they used a core of vitrified or burnt bricks (burrs or clinker), bound and rendered with a cement that was known as Pulhamite. This was coloured to resemble indigenous rocks, and the artificial boulders were skilfully placed, modelled and carved to imitate the disposition and appearance of local natural stone.

They sometimes lined glazed ferneries or caverns with a mixture of real and artificial tufa formed by mixing peat into the render. Wonderful examples of such ferneries were constructed at a house called Acacias in Reading in the 1890s, and at Merrow Grange near Guildford in the 1900s. Probably the

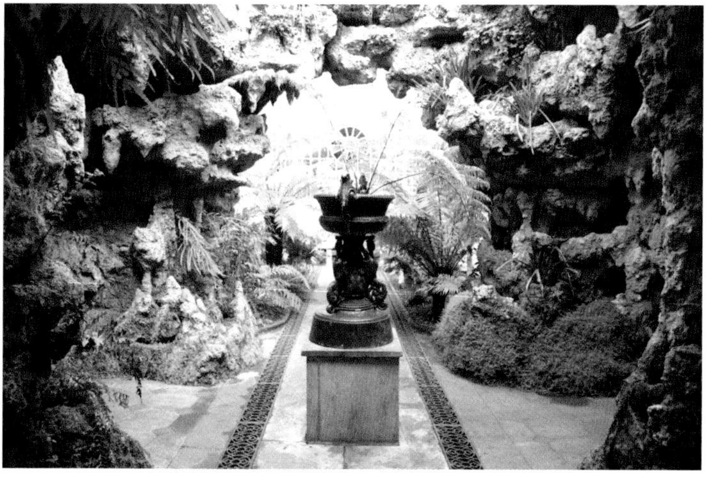

In the 1870s Joseph Shuttleworth converted an 1830s grotto at Old Warden Park into a fantastic grotto/fernery, dripping with Pulhamite tufa forming stalactites and stalagmites and deep planting crevices. (Photo 2009)

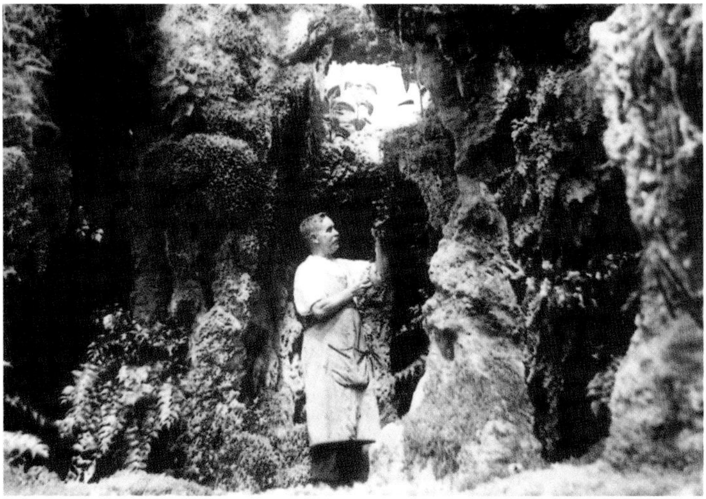

The gardener, Edward Pearce, tends the ferns in one of the Pulhamite grottoes at Dewstow, Monmouthshire. They were begun in the 1890s for Henry Oakley.

largest and most impressive Pulhamite work is at Dewstow Gardens, Monmouthshire, constructed around the same time. This comprises large rock-edged pools, tunnels, and extensive subterranean, tufa-lined grottoes.

If pollution was not too much of a problem native species could be grown outdoors. A small hardy fernery could be made in any shady corner

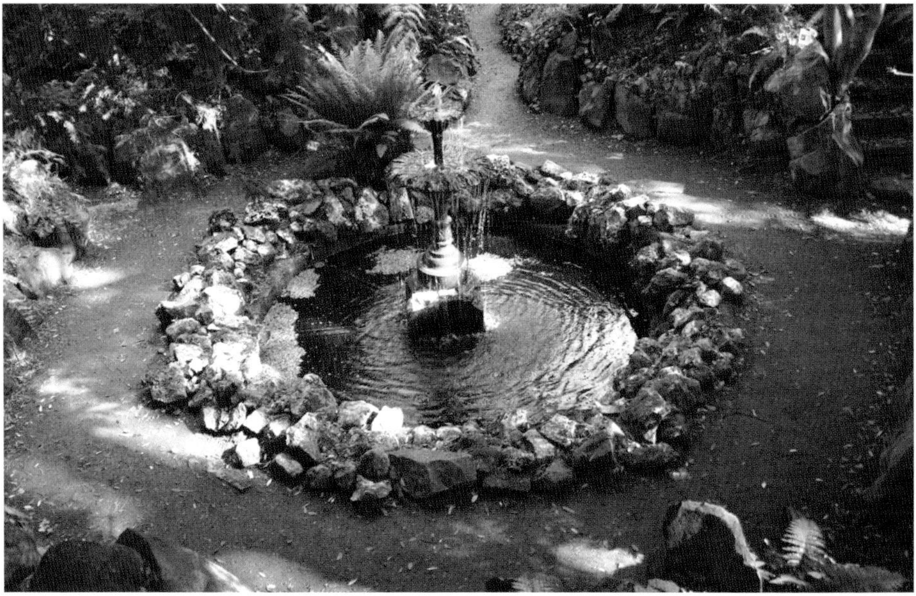

In the 1870s Susannah Harvey remodelled an existing rock garden at Greenway, Devon, into a picturesque fernery. Rustic steps wind up rocky outcrops and serpentine paths lead to a quartz-edged pool and cast iron fountain. (photo 2008)

of a small garden or yard. However, the ideal was a secluded spot some distance from the house, formed in a natural or man-made hollow. Serpentine walks, leading to a rustic seat or summerhouse, were essential. A stream was a great bonus and should be edged with overhanging rocks and crossed by a rustic bridge. Often a rock garden and fern garden were assumed to be one and the same thing. This could vary from little more than low mounds of rock, to rugged outcrops and caves, with water trickling or dripping on to mossy stones.

The Pulhams believed that the work should look as natural as possible, and their rock gardens could comprise huge towering masses of 'boulders' based on geological principles. The aim was to replicate a real stratified rockbed, with bold projections and recesses, fissures, dips and overhangs, all of which provided plenty of pockets for soil and plants together with good drainage. One of their finest outdoor ferneries was at Danesbury Park in Hertfordshire, constructed in a chalk pit in the parkland in 1859–60.

In the 1850s nurseries were advertising exotic tree ferns for sale as stove plants. However, from the 1860s onwards gardeners in parts of

The Pulhamite rockwork fernery at Danesbury Park in Hertfordshire. The gardener, Anthony Parsons, was well known for propagating new varieties of fern.

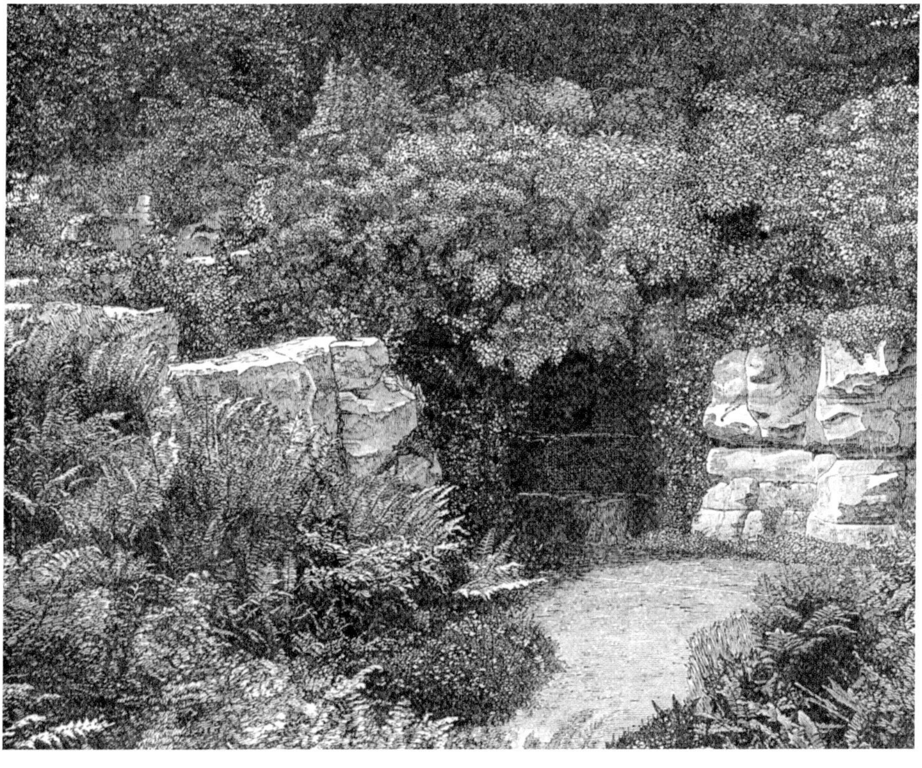

Britain warmed by the Gulf Stream were experimenting with growing *Dicksonia antarctica* in particular outside in the summer, and creating subtropical gardens. The Cornish nurseryman John Garland Treseder introduced many species into the estates of the county. A seemingly unique Cornish feature is the tree fern pit, where specimens were planted in old tin mine or quarry workings.

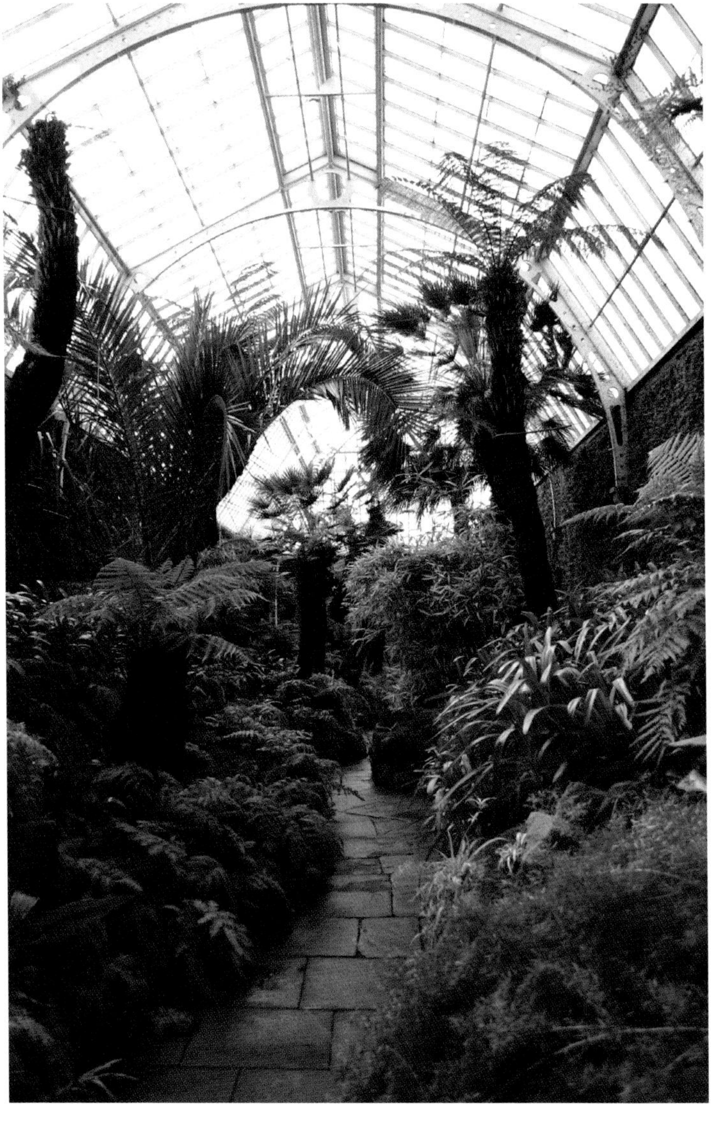

The fernery at Tatton Park in Cheshire. Designed by Joseph Paxton in 1859 to house tree ferns, including *Dicksonia squarrosa* from New Zealand, collected by Captain Charles Randle Egerton in the 1850s. (Photo 2009)

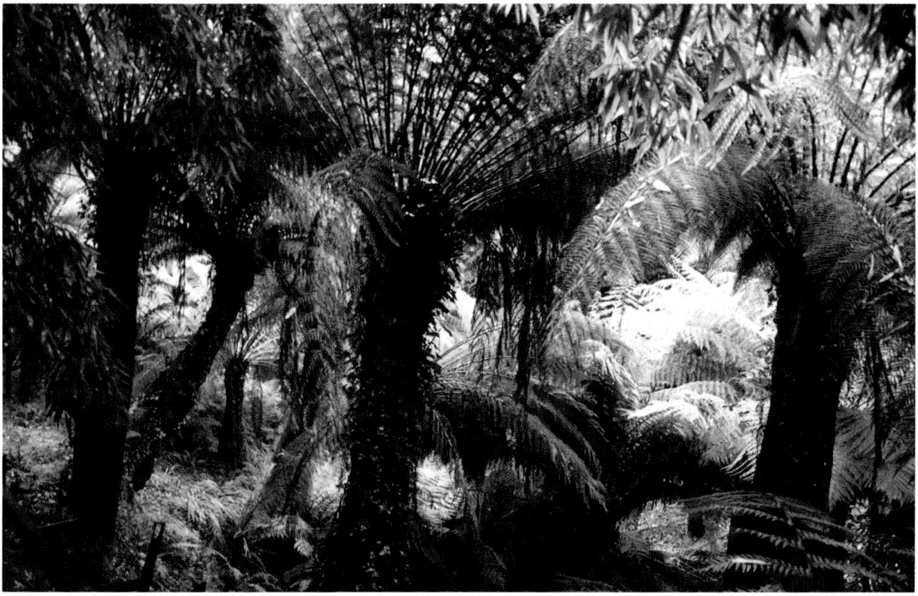

Thomas Bedford Bolitho created a tree fern pit on the site of an ancient open-cast tin mine at Trewidden in Cornwall. It is home to one of the largest collections of *Dicksonia antarctica* in the Northern Hemisphere.

An even more primeval-looking garden feature was the stumpery or rootery, where dead tree stumps were 'planted' upside down so that the roots were in the air. Pieces of contorted wood were also piled up into mounds and arches. This tangled mass was then planted with ferns and other plants such as ivies and hellebores. The best surviving Victorian stumpery was created by James and Maria Bateman, with their friend the fern fanatic Edward Cooke, at Biddulph Grange in Staffordshire.

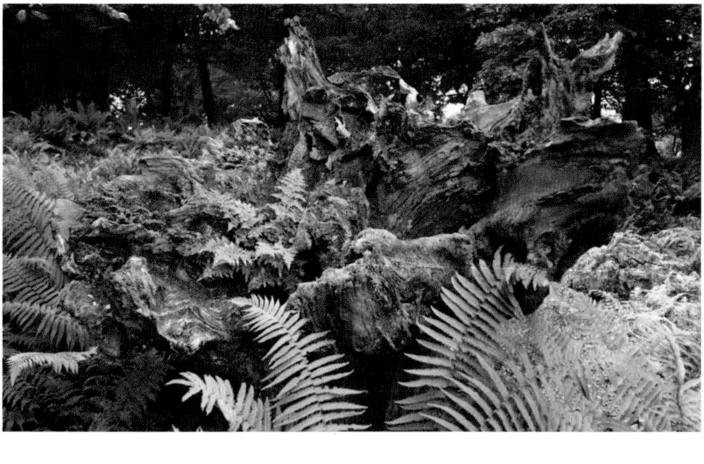

Part of the stumpery at Dyffryn Gardens, Wales. The gnarled and twisted forms of a stumpery are a link back to the Picturesque garden movement. (Photo 2008)

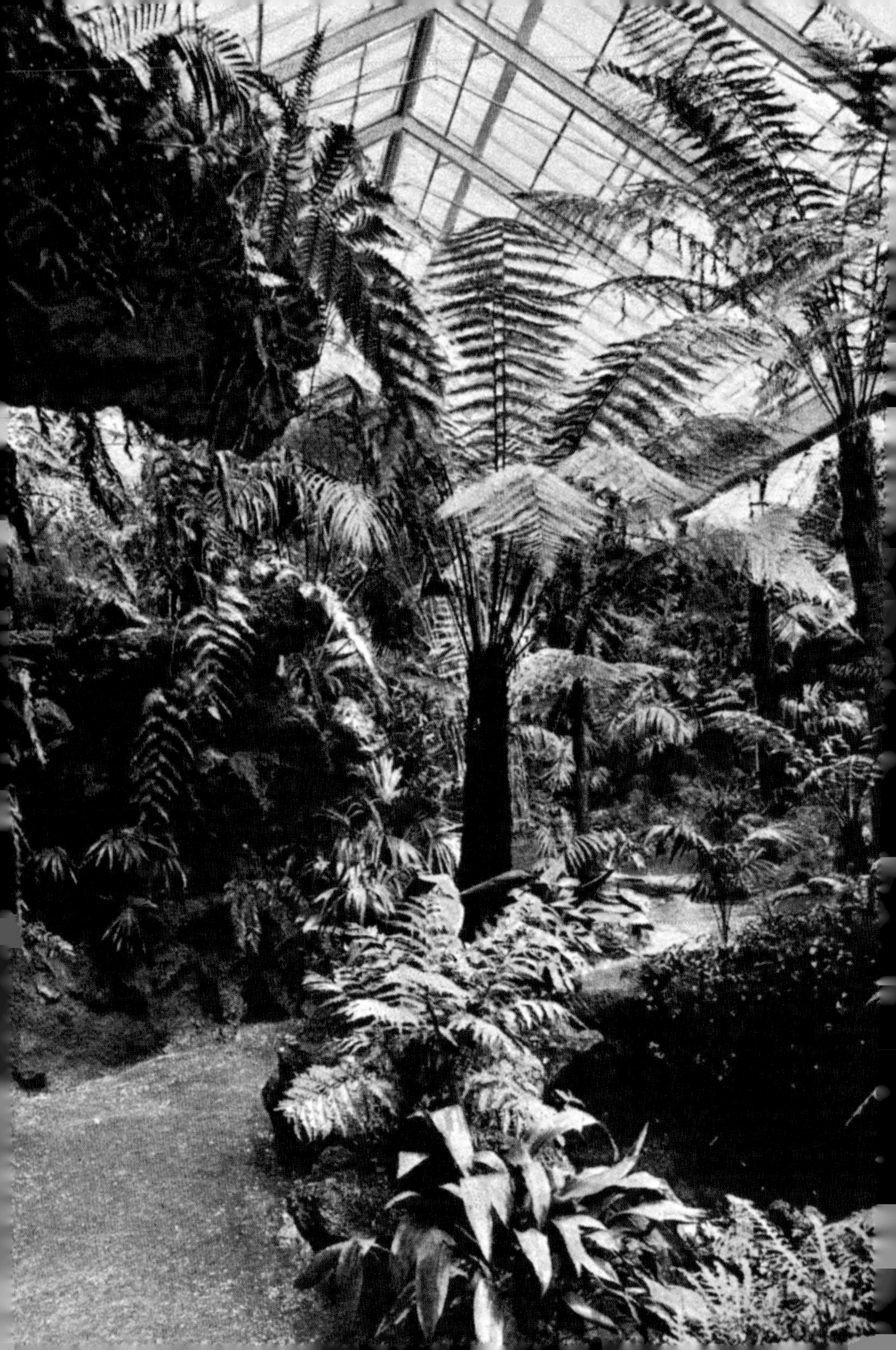

FERNERIES FOR ALL

BOTANIC GARDENS and their curators played a significant role in the fern craze. The gardener John Smith came to Kew Botanic Gardens in 1822, at which time there were about forty hardy species and forty exotic species of fern there. During his time at the gardens (he was curator 1841–64), the collection grew to almost one thousand species, and in 1866 he wrote *Ferns: British and Foreign*. By 1887 Kew possessed a collection of about five thousand ferns (not all different species) displayed in a variety of houses and outside.

The author Thomas Moore trained as a gardener, and in 1841 became assistant editor of the *Gardeners' Chronicle*. He was editor 1866–82, and used the *Chronicle*'s pages to promote fern growing. In 1848 Dr John Lindley secured Moore's appointment as curator of the Botanic Garden of the Society of Apothecaries (renamed Chelsea Botanic Garden in 1875). Moore remained in this position until his death in 1887 and during his term of office oversaw a 50 per cent increase in the number of fern species cultivated there. In the 1860s, with the advice of Nathaniel Ward, he revitalised the garden, including installing experimental Wardian cases.

Some botanic gardens showed off their fern collections in particularly imaginative ways. The Tropical Ravine House in Belfast was described as one of the finest and most artistically arranged fern houses in Europe. It contained a sunken, rocky gully so that the visitor walked around a 'balcony' and looked down on the lush subtropical planting. The gardens at Glasnevin, Dublin, included a special house for the native but rare Killarney fern. In addition to an important tree fern collection in the conservatory, Birmingham Botanical Gardens displayed their hardy species, including *Osmunda regalis*, in the form of a 'Fern Walk' through an excavated sandstone hollow.

Like many other Victorian institutions, botanic and zoological gardens provided a combination of education and recreation, and a fernery was often one element in a cornucopia of attractions. The Southport and Churchtown Botanic Gardens and Museum Company was formed in 1874. Its 20 acres included a museum, boating lake, huge conservatory, and a covered fernery,

Opposite: Fernery at the Royal Botanical Gardens, Old Trafford, Manchester, c. 1909. The gardens were run by the Royal Botanical and Horticultural Society of Manchester and the Northern Counties.

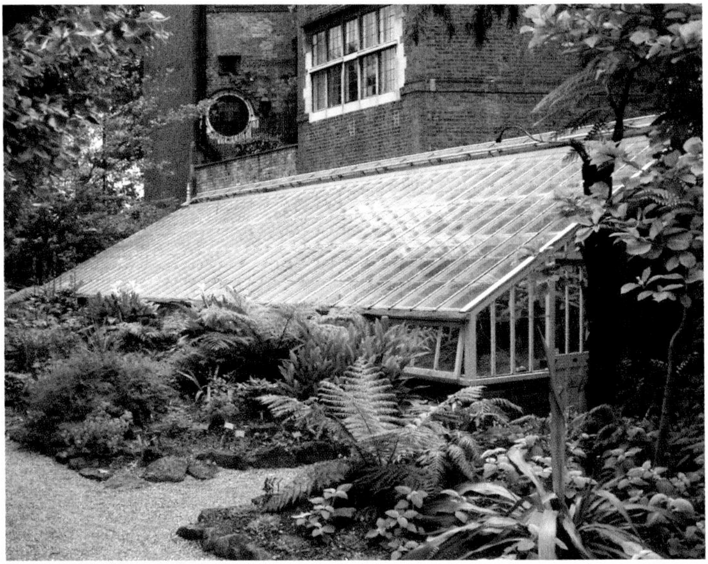

The Thomas Moore Cool Fernery at Chelsea Physic Garden was built in 1907 on the site of the original of 1862–3 (photo 2008).

described as 'some enchanted ravine, where rippling water, splendid ferns, and fantastic rockeries charm the sight.' The Kibble Palace glasshouse was transported from John Kibble's home on the Clyde estuary to Glasgow Botanic Gardens in 1872. Once re-erected and enlarged, a collection of Australasian tree ferns competed for attention against entertainments such as an orchestra performing on an island in a pool.

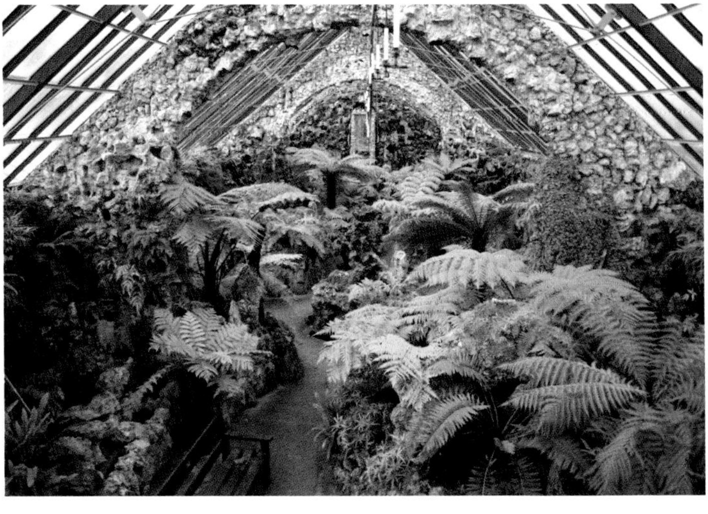

The 1876 fernery at Southport Botanic Gardens. A contemporary guidebook described 'The Unique Fernery With its Fairy-like Foliage, Mirrors, Grottoes, Waterfalls, Fountain and Fishpools, Fine Panoramic Views from the Balcony and Garden Eminences.' (Photo 2009)

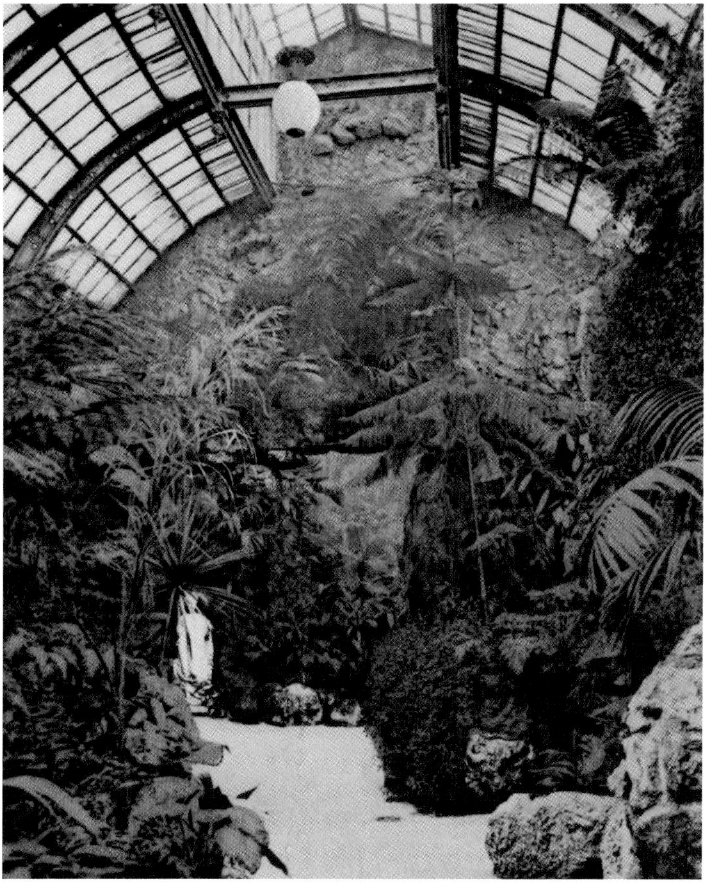

The impressive fernery at Blackpool Winter Gardens. This was just one of the attractions which, over the years, included ornamental grounds, a theatre, refreshment rooms, skating rinks, the Empress Ballroom, 'Indian Lounge', stalactite cave, grotto and Floral Hall.

In the name of 'rational entertainment', public ferneries could be found in an amazing variety of locations. Between 1870 and 1914 over two hundred substantial entertainment buildings were constructed at seaside resorts around the country. The holiday-maker was encouraged to feel removed from day-to-day life, and this was emphasised with oriental-style architecture and winter gardens filled with exotic tropical planting, including palms and tree ferns. In 1889 an advertisement for the vast and magnificent Blackpool Winter Gardens that had opened in 1878 boasted that its grand fernery was 'the finest ever constructed.' It was nearly 200 feet long, with a rockwork wall at one end, waterfalls, paths winding around large boulders, up steps and through arches, and several superb, 30- to 40-foot-high tree ferns.

The parlour aquaria owners soon found that keeping their pets alive was harder than it first appeared, and visiting a public aquarium run by

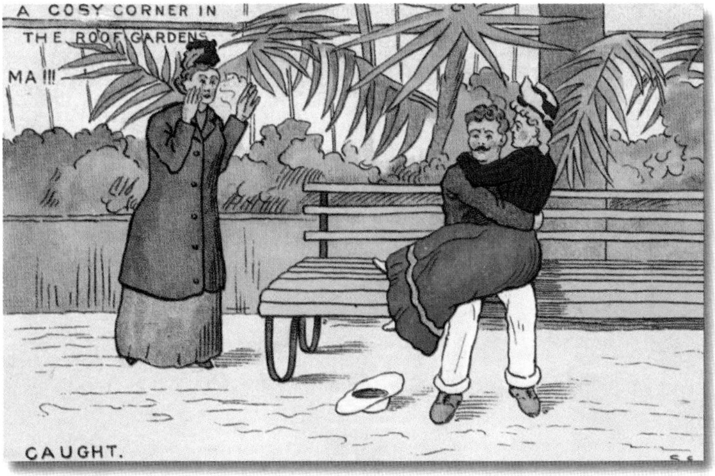

A postcard of Blackpool Tower roof gardens sent in 1906. The lush vegetation, rocky outcrops, and nooks and corners of a fernery provided a perfect location for a romantic rendezvous.

'experts' started to seem more appealing. These were often incorporated into winter gardens, and contained much more than marine creatures. Attractions included spaces for eating, reading, concerts, roller-skating, smoking, and sometimes a fernery. Victorian aquaria were built very much

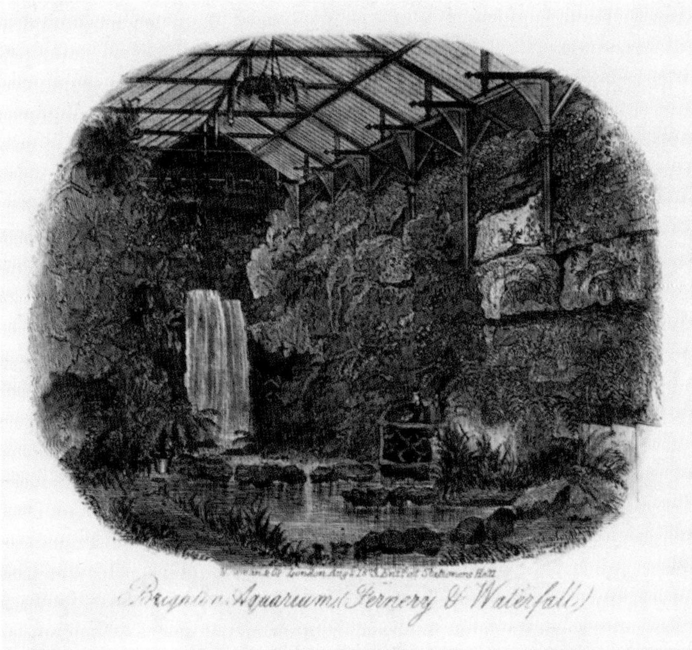

An engraving of 1873 shows visitors standing on a rustic-style viewing platform in the towering Pulhamite fernery at Brighton Aquarium.

in the style of grottoes: they were subterranean, atmospheric places often using tufa or Pulhamite to create a dark, gothic interior. These were suitable conditions for growing ferns, which in their turn added to the green, underwater, ambience.

Brighton Aquarium was designed by the prolific seaside architect Eugenius Birch and opened in 1872. It was constructed between Madeira Drive and Marine Parade next to the Palace Pier, and at its eastern end there was a conservatory and a fernery. This comprised a massive cliff of Pulhamite rockwork with a cascade dropping over the top and forming pools at the base. *The Art Journal* described it as 'one of the most impressive, most effective, and most refreshing sights to be seen anywhere.' The aquarium in the pre-eminent resort of Scarborough was also designed by Birch. It was sited below the Rotunda Museum and the Cliff Bridge, and beneath rows of skylights was a subterranean wonderland of arcades, courts and grottoes in an eclectic mix of eastern architectural styles. The main fernery occupied the central hall, which was 130 feet by 60 feet, lined with tufa, and had a cascade in one corner. A bay off the Entrance Court was earmarked for a second fernery and there were also plans to erect an aviary.

In the early 1900s another seaside fernery could be found at Cunningham's holiday camp for men on the Isle of Man. The island also offered the Fern Glen, which was developed as a tourist attraction in the 1890s, with wooden walkways and rustic bridges to enable visitors to traverse the banks of the picturesque River Groudle. In Dorset, one could eat at the Fern Glen Refreshment Room on the beach at Studland. In north Cornwall there was the Fern Pit Café and Ferry on the River Gannel. Its large garden leading down to the river was full of ferns and was opened as a tea garden in 1910.

Plan of Scarborough Aquarium, showing the positions of the grottoes and fernery.

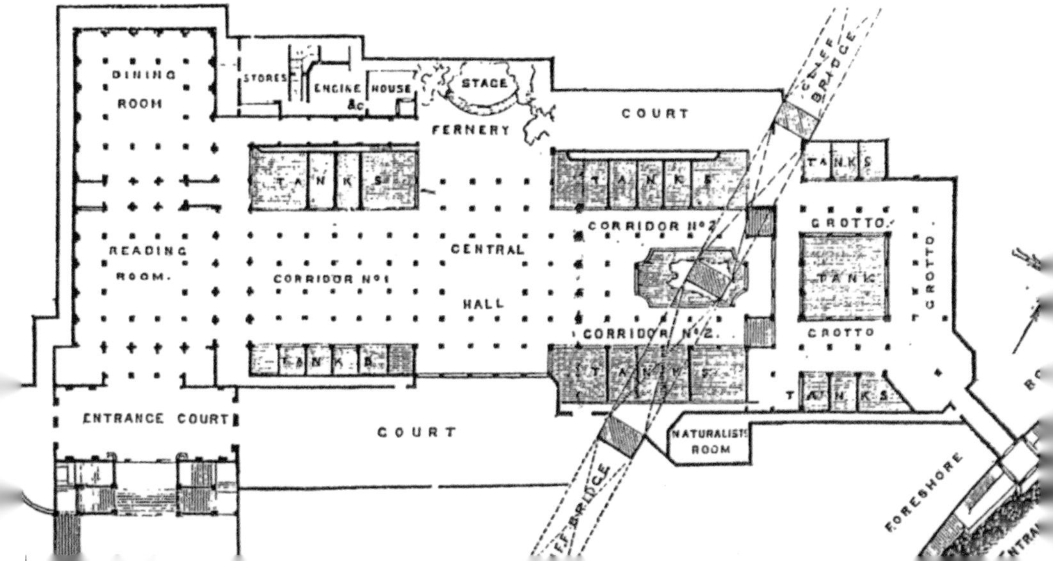

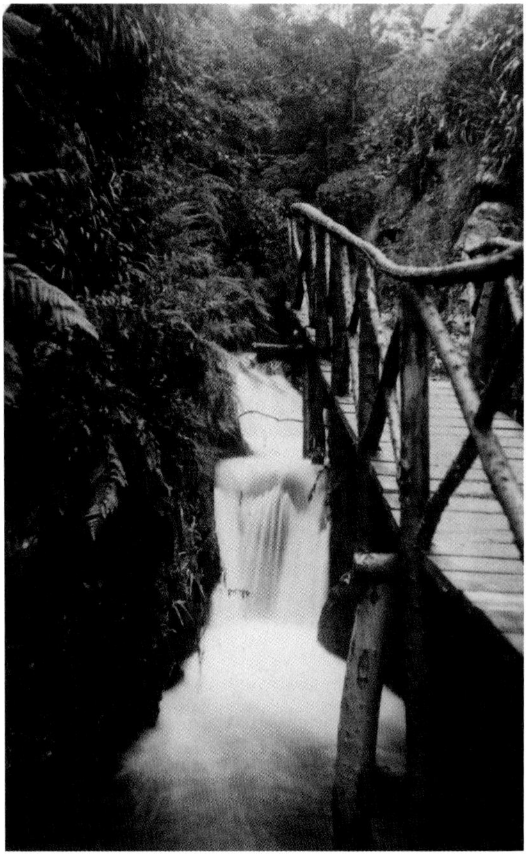

Part of the Fern Glen on the Isle of Man.

When the Cremorne Pleasure Gardens in London re-opened for the 1860 season, new attractions included a picturesque fernery with waterfalls and grottoes. Reached by steamboat, 20 miles down the Thames were the popular Rosherville Pleasure Gardens at Gravesend in Kent, opened in 1837. An old chalk quarry was landscaped into an attractive 20-acre garden, and developed over the years with winding walks, a conservatory, maze, lake, and a fernery. Other amusements included refreshments, dancing, fireworks, a bear pit, theatre and museum.

The extent of the fern craze in mainstream Victorian and Edwardian society can be seen from the number of somewhat surprising places where it was considered appropriate to erect a fernery. It was almost a matter of course to have one at an exhibition, both as one of the exhibits and a place for relaxation and cooling refreshment. The Derby Fine Arts Exhibition of 1877 had ice caverns, stalactite grottoes, rockeries and ferneries which, *The Art Journal* reported, 'added to the attractions and pleased many people.' The new Liverpool YMCA building, opened in 1877, had a fernery off the reading room, and the second-prize entry for a new Conservative Club in the city in 1880 featured a number of 'large elegant ferneries'. When the well-off held dances at home, ferneries would be constructed as cool retreats from the heat of the ballroom.

Ferneries were particularly associated with music and drama. At the Covent Garden Promenade Concerts in the 1870s the processional route was lined with cool grottoes, ferneries and fountains. In 1869 alterations were carried out at the Prince of Wales's Theatre in London, including moving the orchestra to beneath the stage:

> 'an opening by the footlights allowing the sound of the music to be as distinctly heard as heretofore ... The space formerly occupied by the band is now converted into a grotto and fernery, intended, with fountains and

> jets of water, to cool the atmosphere between the acts, and by an ingenious looking-glass arrangement to exhibit an interminably multiplied reflection of tiny crystal rills, which will leap and sparkle in the light through a multitude of leafy labyrinths constructed out of tangled masses of choice ferns most artistically disposed.'

The orchestra members' views on this arrangement are not recorded.

Perhaps most appropriately for a person suffering from Pteridomania, Wardian cases and ferneries were installed in lunatic asylums. Nathaniel Ward believed that cases of ferns and model ruins set into the walls of the long hospital corridors could transform them into 'one of the most beautiful promenades conceivable,' and thereby soothe the afflicted. At Bethlem Hospital in Southwark in the 1860s, ferneries in the form of window cases were fixed on the inside of the corridor windows, 'serving the double purpose of rendering their approach difficult, and adding to the interest of the occupant.'

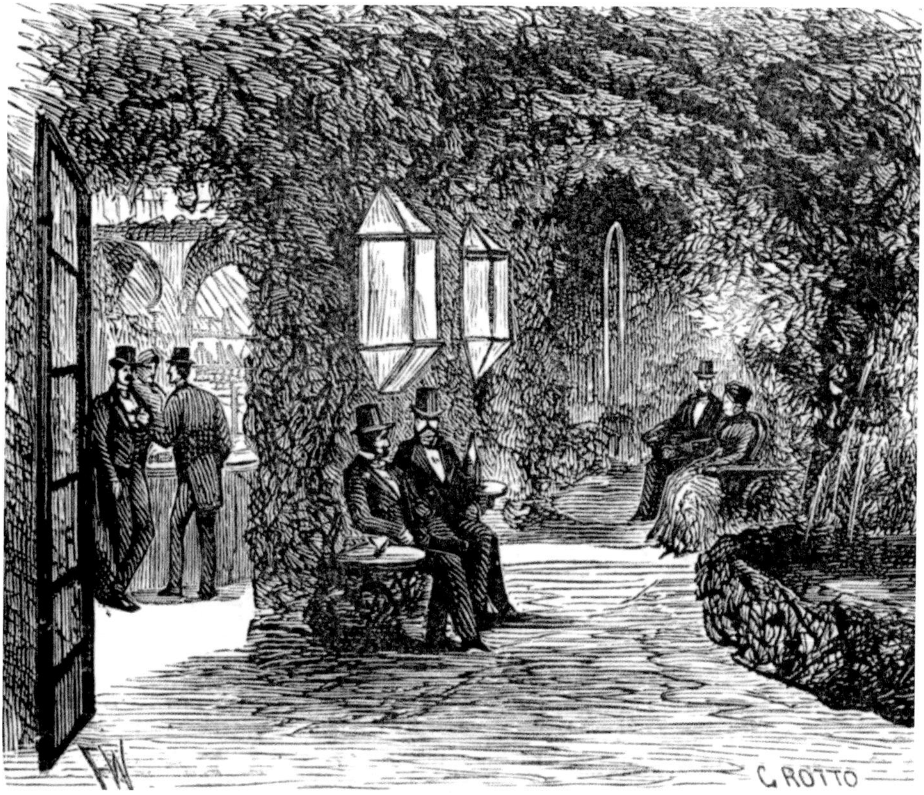

At the newly-built Prince's Theatre in London in 1884, a grotto and fernery extended under the pavement of the street outside.

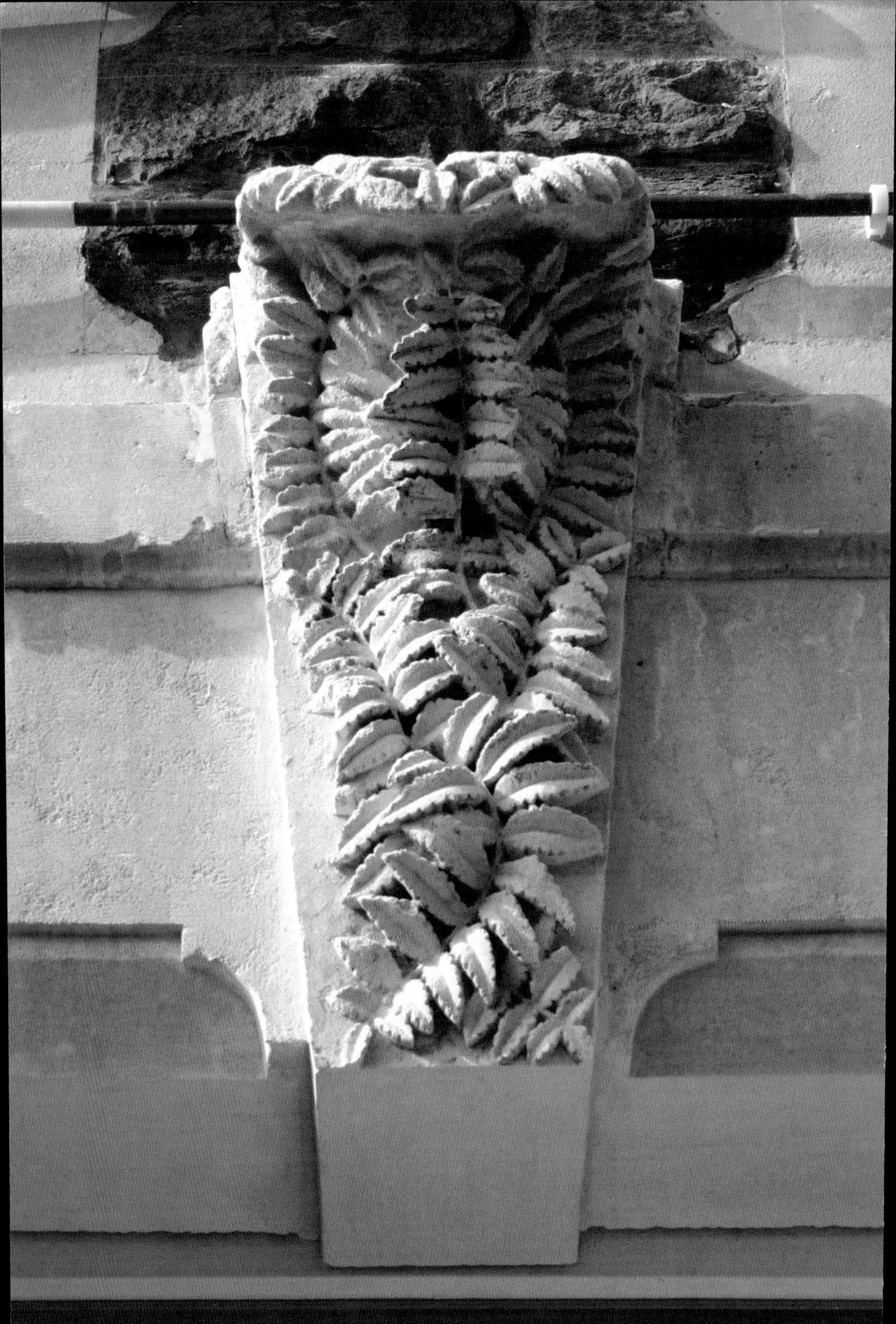

THE DECORATIVE ARTS

FROM THE 1850s onwards there was a widespread use of ornament derived from nature in design and architecture. This, combined with Pteridomania, meant that ferns were frequently depicted in stone, wood and iron on buildings. The fern lover could also fill his or her home with images of the plant, as it appeared on household items from wallpapers to chamber pots. The inspiration of ferns' frondy forms, lacy outlines and dramatic crosiers gave rise to some delightful designs.

Iron was a particularly suitable medium in which to portray ferns. In Australia they were used as a motif in cast iron balcony panels, friezes, and brackets, together with other indigenous plants, animals and birds. In England the Coalbrookdale iron company produced a range of garden furniture in three fern designs *c.* 1860–1900. 'Fern and Blackberry' was made as single chairs, and benches in three different lengths. It was available with a slatted seat made of pine, oak or iron, and came in a brown, green or bronzed finish.

'Osmunda Fern' was a more ornate design, with panels of royal fern on the back, and was offered in a similar range of sizes and finishes. 'Osmunda Regalis' was incredibly intricate, with one large panel of fronds. There was no single seat available in this design, but there was the option of adding small supports on the bench arms to hold drinks. A rustic ferny corner stand with a marble top was illustrated in the Coalbrookdale catalogue for 1875, as were similar umbrella or stick stands, which came in 'green bronzed' or 'Berlin black' finishes.

Ferns were used as a decorative motif in and on a huge variety of media. A 'fern' carpet by Turberville Smith & Co. was exhibited at the Great Exhibition. *The Art Journal Catalogue* proclaimed that the fern was 'one of the most graceful productions of the woods and hedgerows, and worked out in this carpet in shades of the liveliest green, nothing can be more ornamental.' Other textiles, including crochet, damask, printed cloth and ribbons, all featured the fern. The makers of Honiton lace in the area around the town in east Devon frequently depicted birds, insects and plants, including ferns,

Opposite: Ferns and ferny-forms can be found carved in stone all over nineteenth-century buildings, including on capitals, keystones and window surrounds. Roads and houses were also frequently named after the plant.

Below: Honiton lace 'motifs', such as this fern, were appliquéd on to net to create items such as shawls, as well as being sold individually so that ladies could make up their own pieces.

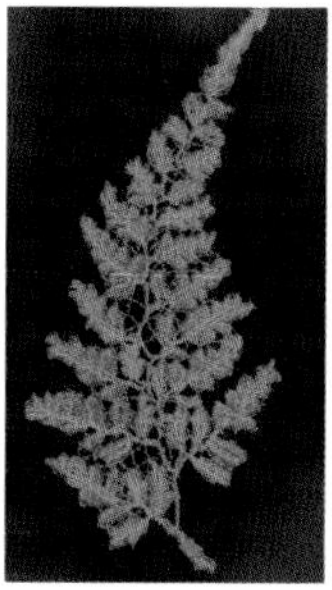

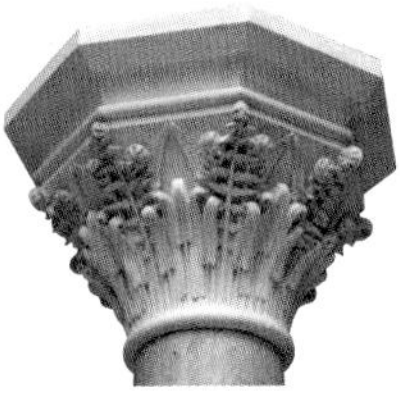

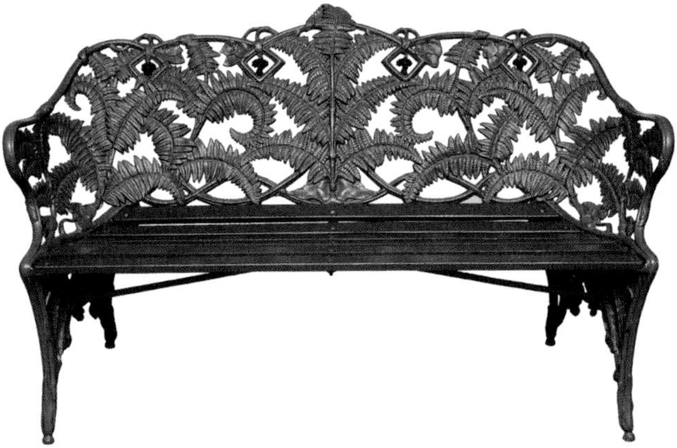

Above: One of the exquisite capitals carved by James O'Shea in the Oxford Museum, 1858–60. The ferns are *Scolopendrium vulgare, Blechnum boreale,* and *Lastraea felix-mas*.

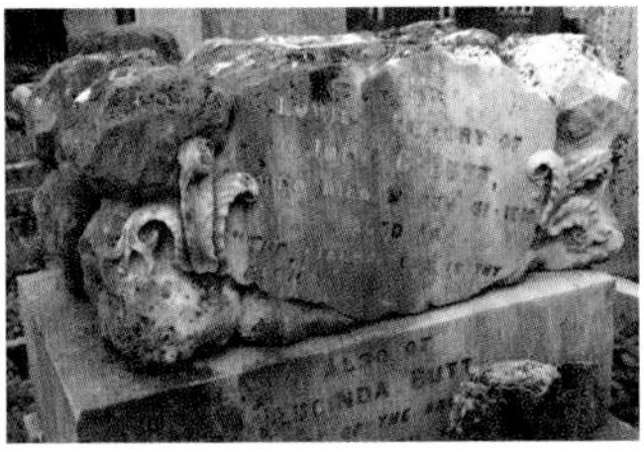

Left: Gravestones were often carved with ferns. This one, in the churchyard at Holy Trinity Church, Ilfracombe, Devon, depicts the forked hart's tongue.

Top right: The Coalbrookdale 'Fern and Blackberry' garden bench depicts *Pteridium aquilinum* or bracken. It was copied by other iron foundries.

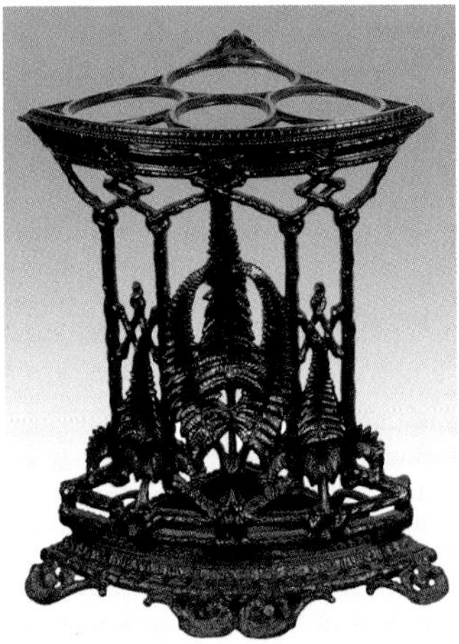

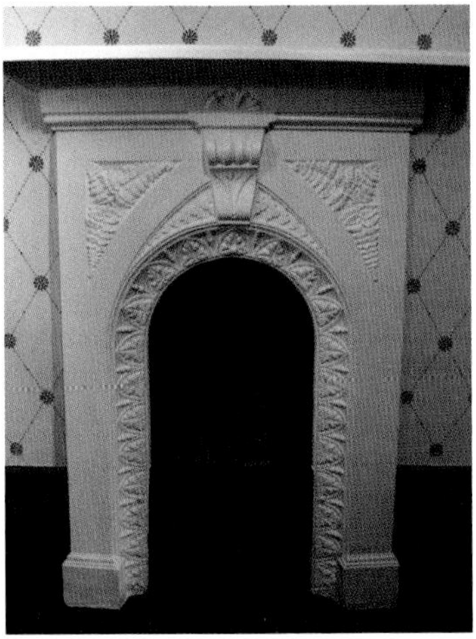

Right: A Coalbrookdale cast iron umbrella or stick stand with a fern design. It was illustrated in the company's 1875 catalogue.

Far right: A cast iron fire grate with ferns in the spandrels. The design was registered between 9 December 1870 and 3 May 1871.

that they could find in the countryside. The lace was very fine and delicate, and the best work was so realistic that it is possible to recognise the species represented. One surviving piece illustrates the underside of a male fern frond, including the spores, while another is reported to have portrayed the whole fern, root and all.

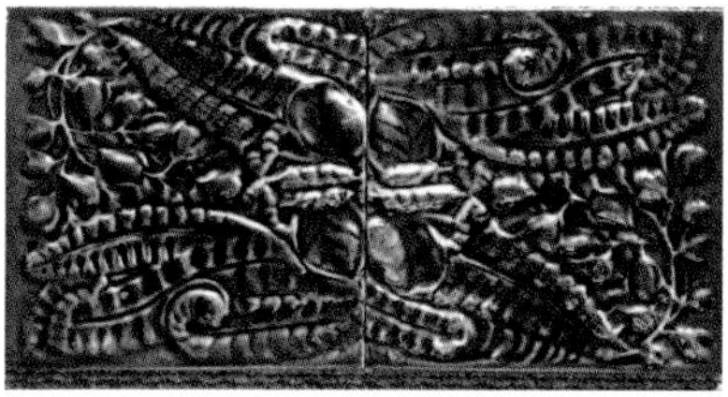

These beautiful moulded and glazed tiles depict a number of ferns, including a variety of hart's tongue, *Asplenium scolopendrium undulatum*, with a ruffled edge.

Perhaps the widest array of fern designs is to be found in the field of ceramics. Most of the major British potteries produced some, including Minton. A set of tiles featuring ferns and other woodland plants was offered by a tile-making side of the business, Minton, Hollins & Co., in 1876. Among various fern patterns, Wedgwood made a set of transfer-decorated plates using illustrations from Anne Pratt's 1855 book, *The Ferns of Great Britain and Their Allies*. In the 1880s Ridgways of Stoke on Trent designed an earthenware dinner service with a pretty, delicate transfer design called 'Maiden Hair Fern', which came in reddish brown, inky blue or dark green on cream.

One of the most natural-looking and attractive designs was the range of porcelain jugs and sugar bowls made by Royal Worcester from the 1890s onwards. They were formed of overlapping hart's tongue fronds, a taller one creating the lip of the jug and a separate one the handle. They vary enormously in size from about 3 to 8 inches high, and are usually found in a glazed, creamy-white finish. However, some jugs and larger bowls were

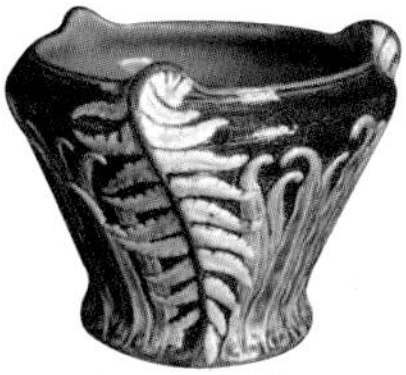

Above: Majolica, with its deep colours and reflective glazes, was a particularly suitable technique in which to portray the fern. Both Minton and Wedgwood made such pieces.

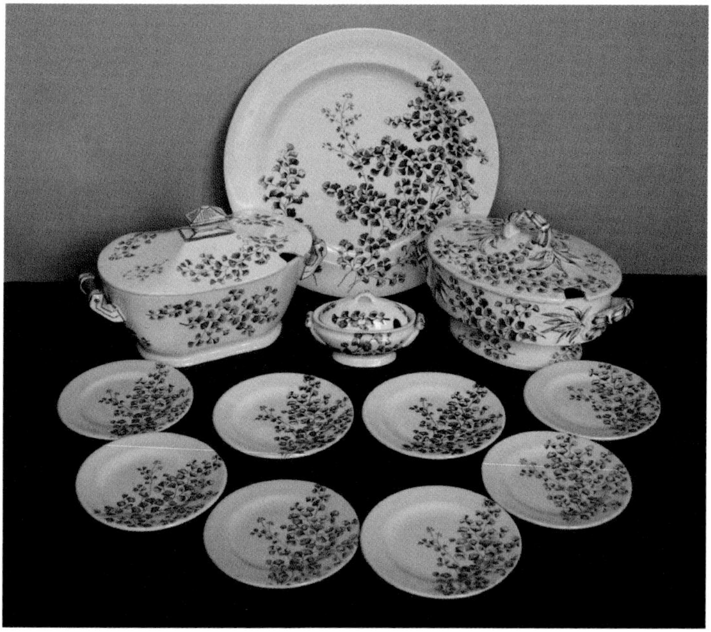

As well as full-size dinner services, Ridgways made children's toy tea and dinner sets in their 'Maiden Hair Fern' design.

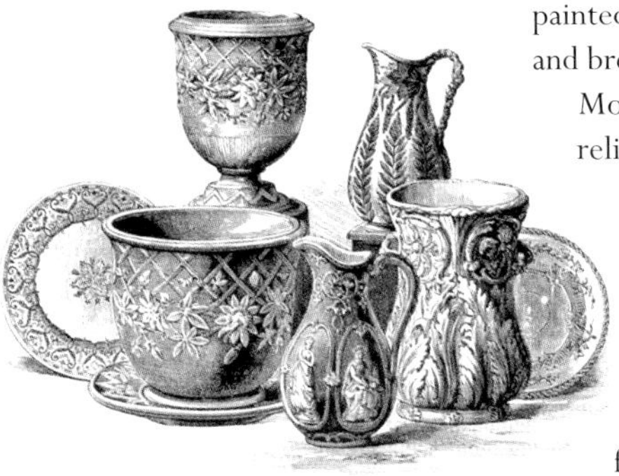

Right: William Brownfield exhibited its 'Fern' design at the London International Exhibition of 1862, and it was illustrated in the catalogue (back right). Although stylised, the fronds on the jug were actually more ferny in reality than depicted here.

painted in soft tones of green, yellow and brown, with a gilt edge.

More robust in style were the relief-moulded jugs produced by a number of companies. The Staffordshire pottery William Brownfield registered a stoneware jug design called 'Fern' on 5 November 1859. It consisted of stylised, erect fern fronds with heads of wheat hanging down between them from an acanthus leaf border. It also had a 'rope' handle with a 'knot' in it. This was its most popular jug design and was produced in at least sixteen different colour combinations, including one in majolica. It was so successful that it was quickly, and illegally, copied by numerous other firms, including Samuel Bevington. James Dudson actually registered a very similar design on 27 February 1862, although his pattern is more like acanthus leaves than fern fronds, and there are no ears of wheat. When the three-year protection of Brownfield's registered design expired, variations of it were registered by Worthington & Green, James Macintyre, F. R. Pratt, and Wood & Sale.

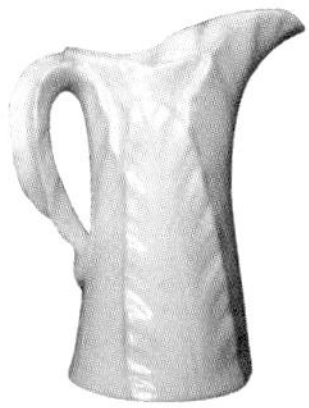

Above: A Jug based on the hart's tongue fern, produced by the Royal Worcester company from the 1890s to 1950s. The design was revived in the 1980s when a gilt edge was added to the lip of the jug.

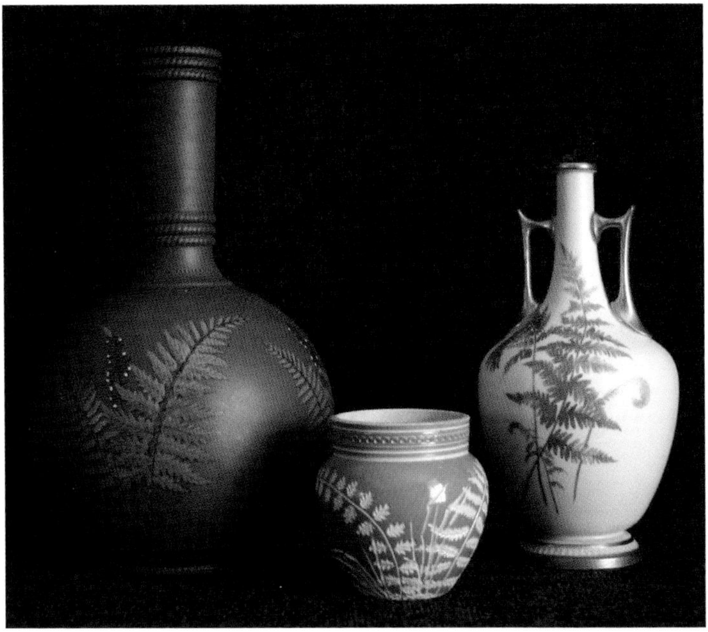

Right: A selection of fern ceramics, left to right: a terracotta vase; a Salopian Art Pottery Company glazed pot with incised decoration; a Royal Worcester gold enamelled bud vase, dated 1889.

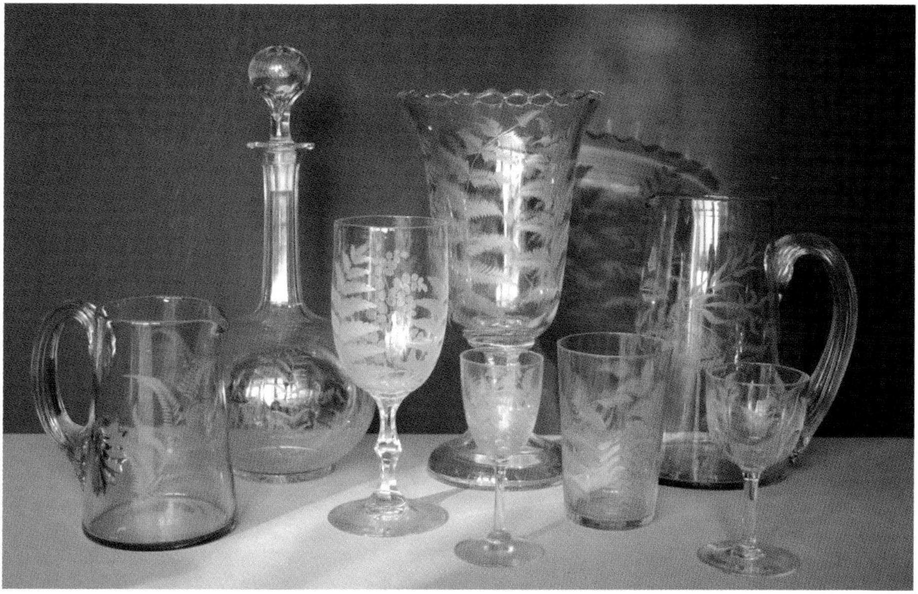

Glass decanters, flower and celery vases, water jugs and drinking vessels were frequently decorated with ferns in the Victorian period – some with recognisable species, others with very sketchy fern-like plants.

From the 1850s through to the 1900s, the Dudson factory also produced a wide range of Jasperware decorated with fronds. Cheese dishes, pots and jugs came in various colours with applied white 'sprigging'. Designs included 'Ferns', 'Fern Wreath' and 'Bluebell and Fern'. Art potteries also created fern designs: the Watcombe Terracotta Company sold items decorated with enamelled ferns, and the Salopian Art Pottery produced a wide range of earthenware with incised fern decoration from the 1880s onwards.

Manufacturers catered for all pockets, and numerous inexpensive, unmarked ferny pieces can still be found. The same is true of glass items, to which designs were applied by engraving and other techniques. The results were frequently very attractive and the fronds accurately represented. Often, however, particularly at the cheaper end of the market, they were not. Indeed, many surviving examples are decorated with crude foliate patterns that look as though they were probably scratched on at home.

At the opposite end of the scale, some of the highest quality glass was made and engraved in Scotland, by companies such as the Bathgate Glass

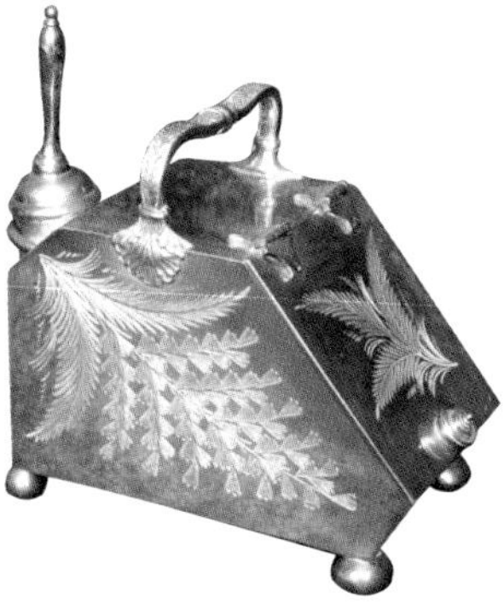

As with glass, the quality of engraving on silver varied. This 'sugar scuttle' made by James Dixon & Sons of Sheffield in their Electro Plated Britannia Metal (EPBM), sometime after 1879, is engraved with stylised maidenhair ferns and palms.

Works (1866–87). Scottish glass engraved with ferns at the workshop of J. H. B. Millar in Edinburgh was displayed at the 1862 International Exhibition, when it was said that 'especial study has been paid to render the ornamentation in harmony with the forms.'

Sometimes a piece of Mauchline Ware contained a glass engraved with ferns. This range of wooden tourist souvenirs was produced in south-west Scotland from the early nineteenth century until the 1930s. It is an indication of the intensity of Pteridomania that in addition to Tartan Ware, and Transfer or Photographic Ware depicting topographical views, a category called Fern Ware was made from about 1870. A huge range of objects was produced and decorated with brown fern-printed paper, coloured transfers or reduced photographs of ferns. If the item was large enough, dark brown ink was stippled over pressed and dried layers of fronds pinned directly onto it. All pieces were finished off with many layers of varnish.

In New Zealand some souvenir fern albums had wooden covers with carving or marquetry accurately portraying indigenous species. The best were created by Anton and William Seuffert, who also made stunning inlaid furniture. Similar pieces, known as Killarney Ware, were made in Ireland for the tourist trade. They were decorated with fine marquetry work depicting local scenes, traditional Irish symbols and native ferns.

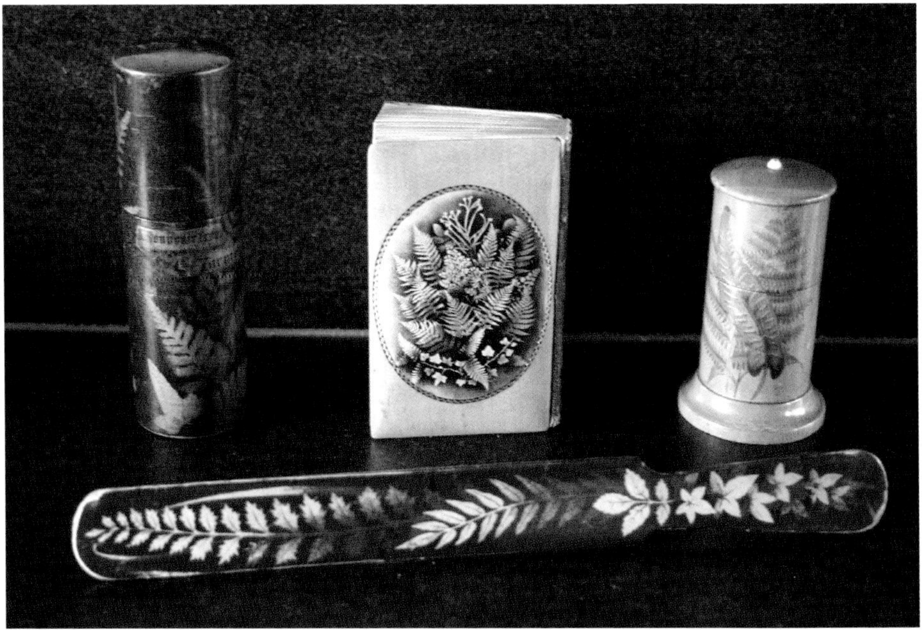

Examples of Mauchline Fern Ware: a 'Souvenir from Yarmouth' with imitation spatter-work; a letter opener covered with printed paper; a miniature book with covers decorated with photographs of ferns surrounded by typical gold, wavy-line borders; a container with coloured transfers.

Fern lovers could wear ferns in the form of jewellery, made of silver, Whitby jet, bog oak or pressed horn, and send each other birthday, Christmas, and new year greetings with cards designed just for them. Young ladies also made their own ferny items. They compiled albums and framed arrangements of pressed ferns, not necessarily as part of a botanical collection but for aesthetic effect. They also used dried ferns to decorate small items such as boxes, or inked them to make prints. The most popular pastime was 'spatter-work', when pressed fronds pinned on to pieces of paper, fabric or pieces of furniture were covered with a spattering of India ink from a tooth-brush

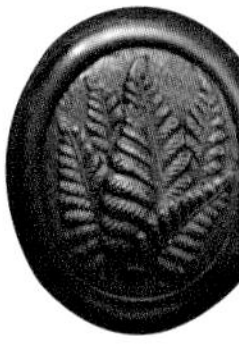

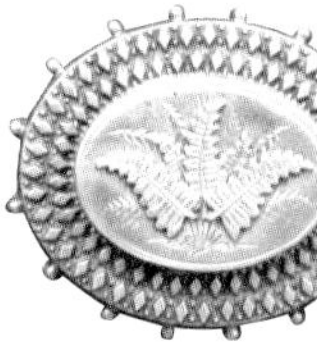

Top: a brooch made of moulded horn dyed black, with ferns in relief.

Above: a silver-plate brooch with ferns engraved and in relief. It has a compartment to hold a picture or lock of hair. Both could have been mourning pieces.

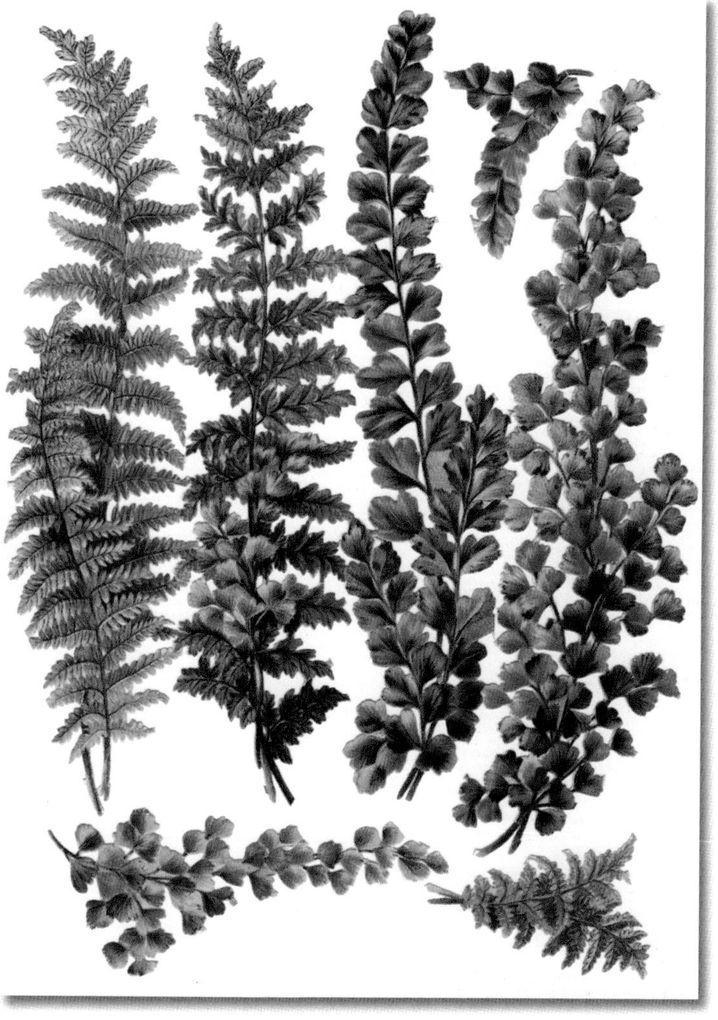

Die-cut, chromolithographic stamped, embossed fern scraps, from Raphael Tuck & Sons's 'Mixed Ferns' sheet, late 1890s. Scraps like these, sometimes mixed with real pressed ferns, were pasted into albums.

Various fern greetings cards. The card top left, posted in 1906, has a real piece of maidenhair fern attached to it.

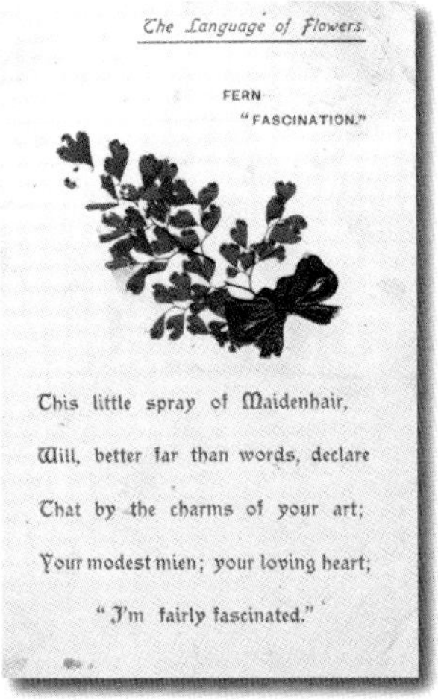

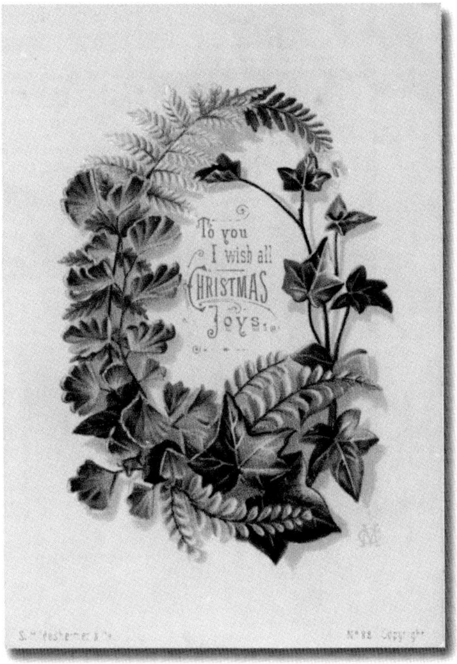

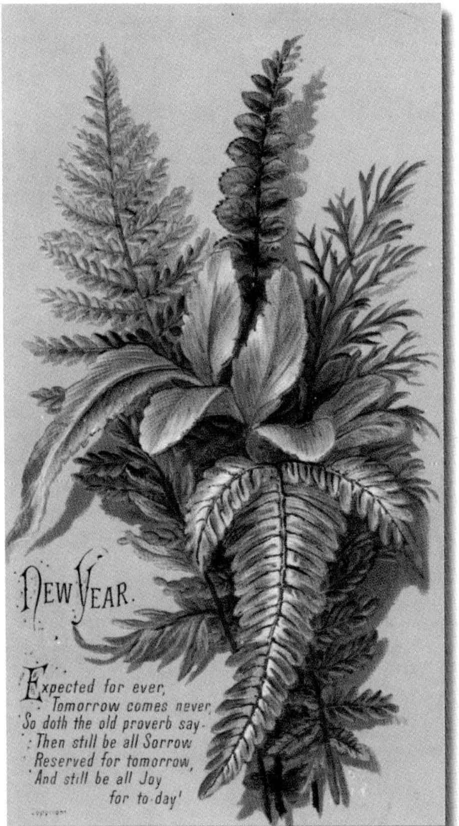

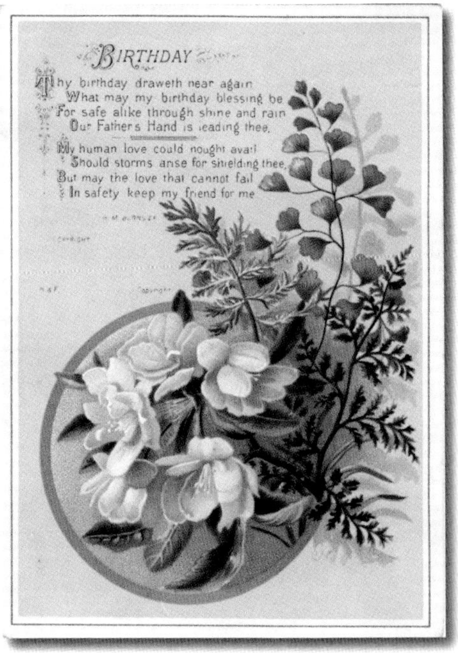

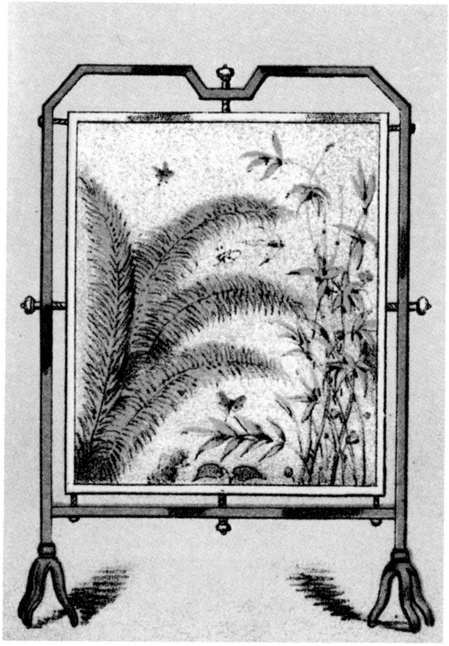

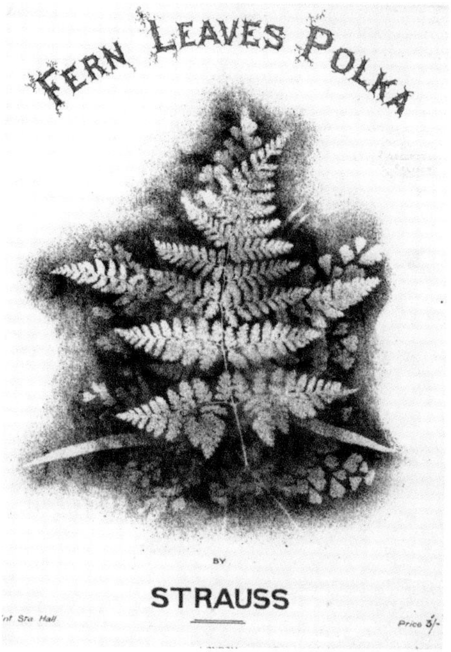

Above Left: Mrs Julia McNair Wright provided instructions for creating a fire screen with pressed ferns held between two panes of glass in *The Complete Home: An Encyclopaedia of Domestic Life and Affairs* (Philadelphia, 1883).

Above right: Fern lovers could even surround themselves with the 'music of ferns', including the 'Beautiful Ferns Valse', the 'Fern Leaves Polka', and the 'Golden Ferns Polka'.

drawn across a comb or flicked through a wire sieve. Very delicate and detailed silhouettes were revealed when the ferns were removed.

Ferns were one of the most popular plants to use in the production of 'phantom bouquets', which were fashionable with American women in the 1860s. Foliage was macerated (treated to remove its fleshy parts) and the resultant skeleton bleached and then arranged in white, ethereal bouquets, often under glass domes. Their transient and ephemeral appearance made phantom bouquets particularly suitable for framing portraits of deceased family members or national figures in memorial stereographs.

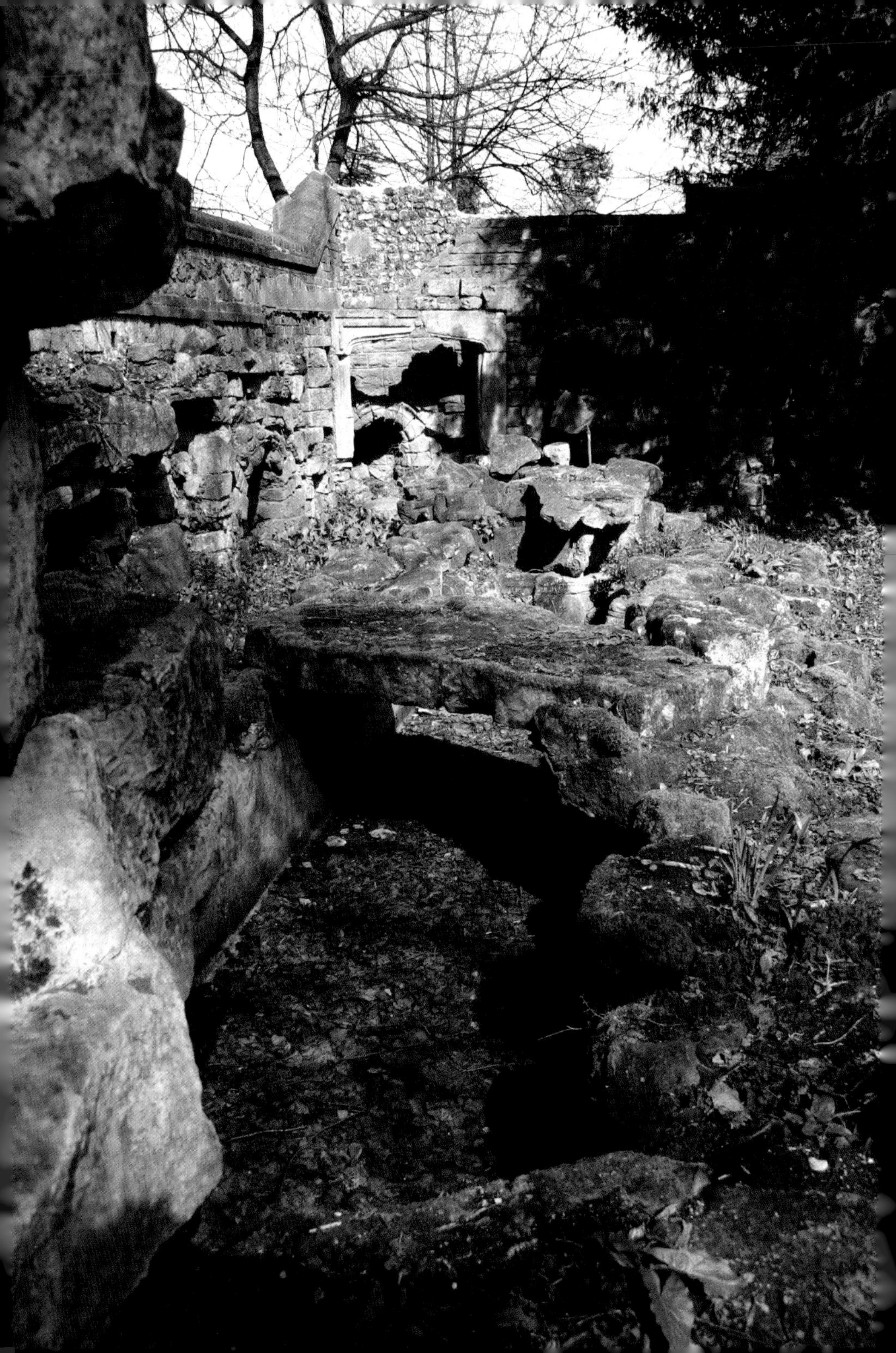

THE END OF THE CRAZE?

Of course the acute intensity of Pteridomania faded over time, but the public's love affair with the fern frond continued into the Edwardian period. A good indication of this is that fern touts still found a market for their wares in the 1900s. The *Gardeners' Chronicle* reported in 1904 that three people had been convicted at Axminster in Devon of stealing 3 cwt of ferns from a private estate. And in 1906 a bye-law was introduced in Devon and Somerset against uprooting ferns and other wild plants growing in any public place.

Although many of the famous names in pteridology had died by this time, they were replaced by new fern breeders such as T. E. Henwood, whose polypodium collection was famous in the early twentieth century. Most of the main nurseries were still extant. In 1907 the Birkenhead stock was acquired by H. B. May & Sons, who brought out the third edition of Birkenhead's book, *Ferns and Fern Culture,* in 1911. Before the First World War May's was considered the pre-eminent fern nursery in Britain, its major rival being James Hill & Sons of Lower Edmonton.

Books also continued to appear. Edward Step's *Wayside and Woodland Ferns* was first published in 1908 and reprinted six times. Other titles included Alfred Hemsley's *The Book of Fern Culture* (1908), and S. Leonard Bastin's *How To Know The Ferns* (1917). Until his death in 1913, Francis George Heath continued to find new aspects of the subject on which to write. In 1906 he proposed in *The Times* that businessmen could erect ferneries in the 'dismal forecourts and rear quadrangles of public offices', which would transform them into 'dreams of beauty'. Pteridology had always been considered a wholesome pursuit for young people, and in 1910 Heath wrote *Fairy Plants: A Fern-book for Children.* In 1903 Charles Druery was still recommending his readers to try fern hunting, hoping that their success would 'inoculate them with that lasting and incurable disease known as the "fern fever".' His last book, *British Ferns and Their Varieties,* was published in 1910 and he wrote articles on fern variety hunting until his death in 1917.

Opposite:
A ruined, roofless fernery in Hampshire awaits restoration in 2009. Although some authors advocated the employment of 'architectural salvage' as a backdrop to the planting, this fernery is a rare surviving example of such a use.

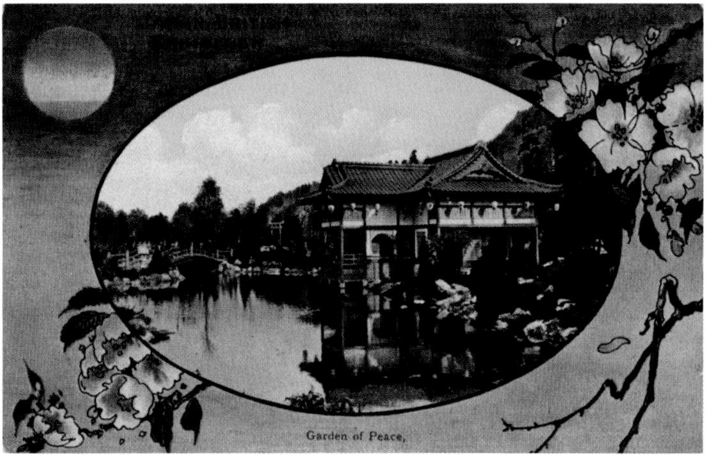

Right: The Japan-British Exhibition held in London in 1910 helped popularise a Japanese-style of gardening in Edwardian Britain.

Below right: The cover of Daniel Ferguson's *British Ferns, Clubmosses, & Horsetails,* published in 1912.

Below: Part of H. B. May & Sons's nursery, from Alfred Hemsley, *The Book of Fern Culture,* published in 1908.

Inevitably, however, fashions in gardening had changed. One way in which ferns survived the arrival of new horticultural tastes was their inclusion, along with bamboos and mosses, in Japanese-style gardens. The fern expert Edward Lowe lived at Shirenewton Hall near Chepstow from 1880, and after his death in 1900 the house

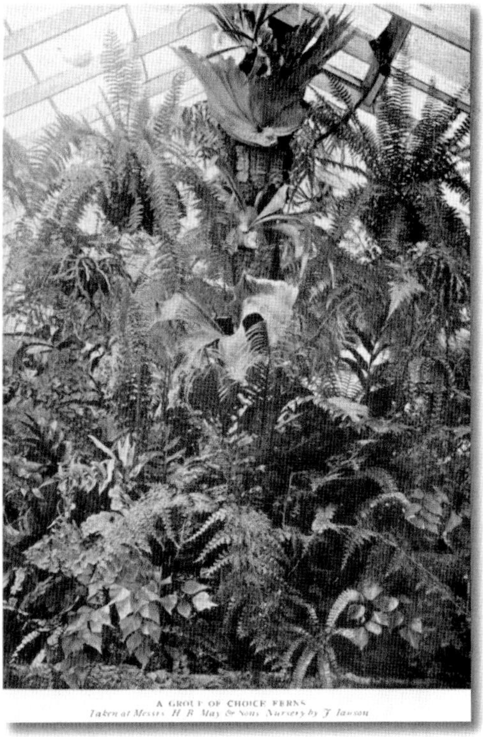

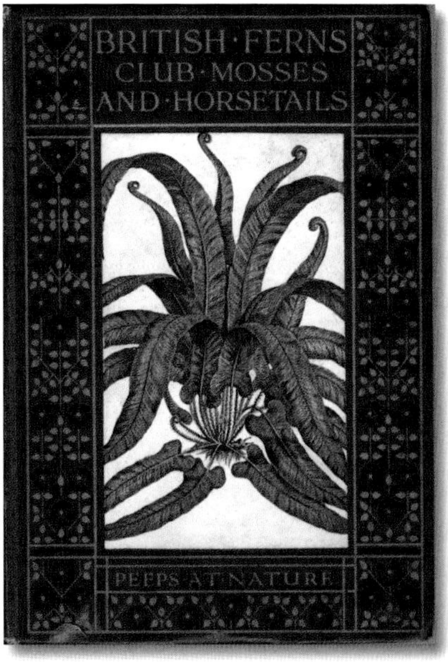

was bought by Captain Charles Liddell. He was a wealthy shipper who traded in the Far East and was therefore able to import authentic structures, including a pagoda, for his new Japanese garden. This featured ponds, cascades, maples and pine trees, together with many of Lowe's ferns.

The rock garden was also extremely popular in the Edwardian period and 1920s. By this time it was synonymous with alpines and had become 'the rockery'. However, ferns were still considered in keeping with rocky landscapes, and existing collections were sometimes incorporated. For example, the fern collection at Sizergh Castle in Cumbria was started in the Victorian period, and when the rock garden was constructed there in 1926 the ferns were moved to it.

But every craze comes to an end, and Pteridomania was no exception. The start of the twentieth century saw a move out of the hot house into the fresh air, and the loggia became a more familiar feature in the garden than the stuffy conservatory. Wardian cases also went out of fashion and what had been a 'must have' in the 1860s inevitably ended up on the rubbish heap. Many people had hoped that they were totally maintenance free, which of course they couldn't be, but in addition the more widespread use of electric light in place of gas removed the need for a protective glass case for plants. But, as with so many things, it was really the Great War that finished the fern craze; garden staff and knowledge were lost, and the ferneries quietly subsided and collapsed.

Ferns were probably at their lowest level of popularity between 1914 and 1945. However, in the 1950s they were already making a comeback,

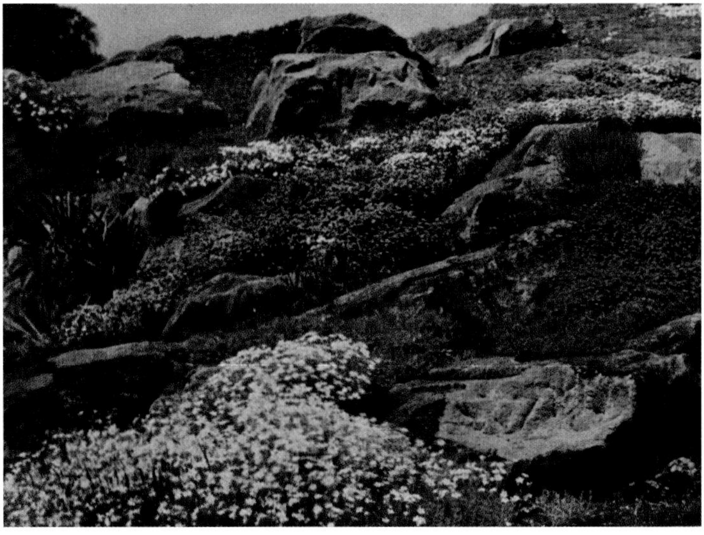

Rockeries were synonymous with alpines in the Edwardian period and later, but ferns were often still allocated space (illustration from E. H. Jenkins's *The Small Rock Garden*, 1913).

perhaps as part of the rediscovery of the British countryside after the end of the Second World War. *The Observer's Book of British Ferns* was published in 1950 and was soon joined by other titles such as George Rees's *How to Grow Your Own Ferns Indoors*. In the 1960s and 70s bottle gardens were all the rage, reintroducing many to ferns' appeal. But it was in the 1990s that the plant, in the form of tree ferns, enjoyed a huge revival in popularity in Britain. Perhaps because global warming had increased their survival rates, they became *the* plant to grow in the garden, leading to a wider appreciation of ferns and ferneries generally.

From the mid 1990s Victorian and Edwardian ferneries and Pulhamite gardens – including Ascog, Dewstow and most recently Benmore, to name just a few – started to be restored and new ones created. The Eden Project, opened in a former china clay pit in Cornwall in 2001, contains the biggest conservatories in the world in the shape of two vast 'biomes'. They evoke the spirit of the great exhibitions of the nineteenth century that brought the world's flora and fauna under one roof, and receive equally vast numbers of visitors. The British public may not have fully succumbed to Pteridomania again just yet, but they are definitely displaying symptoms.

After standing derelict for over a century, the fernery at Benmore was fully restored 2008–9. This huge (1,532 square feet) glazed and heated structure was constructed for the sugar refiner James Duncan in the 1870s. Inside it was on three levels with a grotto (photo 2008).

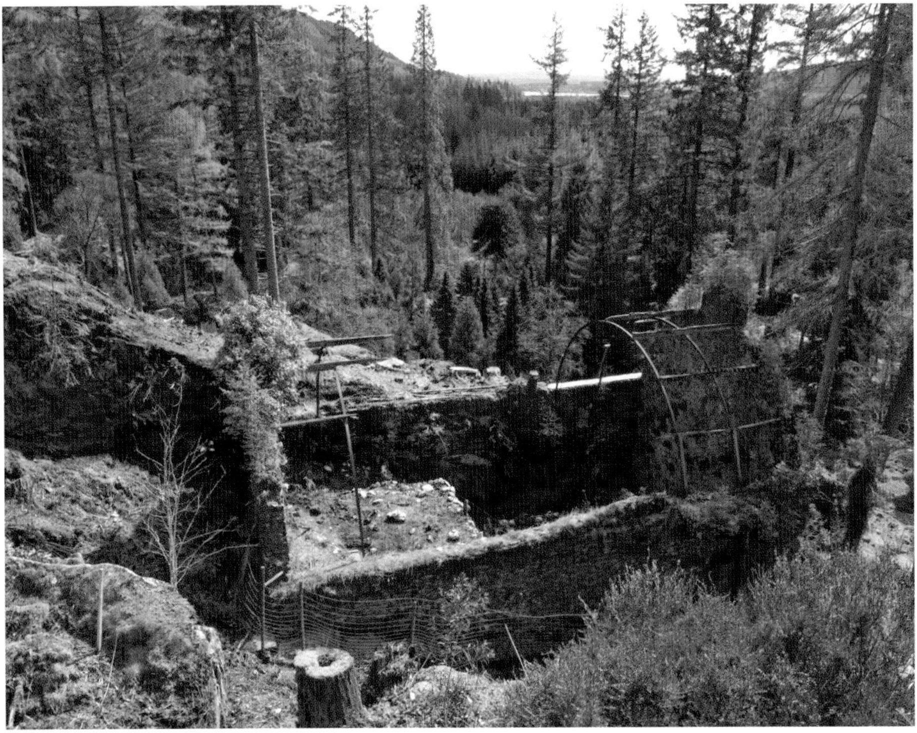

FURTHER READING

For information on all things fern-related, see the British Pteridological Society's (BPS) web site: www.nhm.ac.uk/hosted_sites/bps

The BPS *Bulletin*, *Fern Gazette*, *Pteridologist*, and Special Publications.

Claude Hitching's site, 'The Pulham Legacy' at www.pulham.org.uk

Hall, Nigel, and Martin Rickard. *Fern Books in English Published Before 1900*. London: BPS Special Publication no 9, 2006.

Rickard, Martin. *The Plantfinder's Guide to Garden Ferns*. Newton Abbot: David & Charles, 2000.

The rainforest biome at the Eden Project contains a huge range of exotic planting, including various species of fern such as *Cyathea* and *Platycerium*, together with *Davallia canariensis*, *Diplazium esculentum*, and *Stenochlaena palustris*.

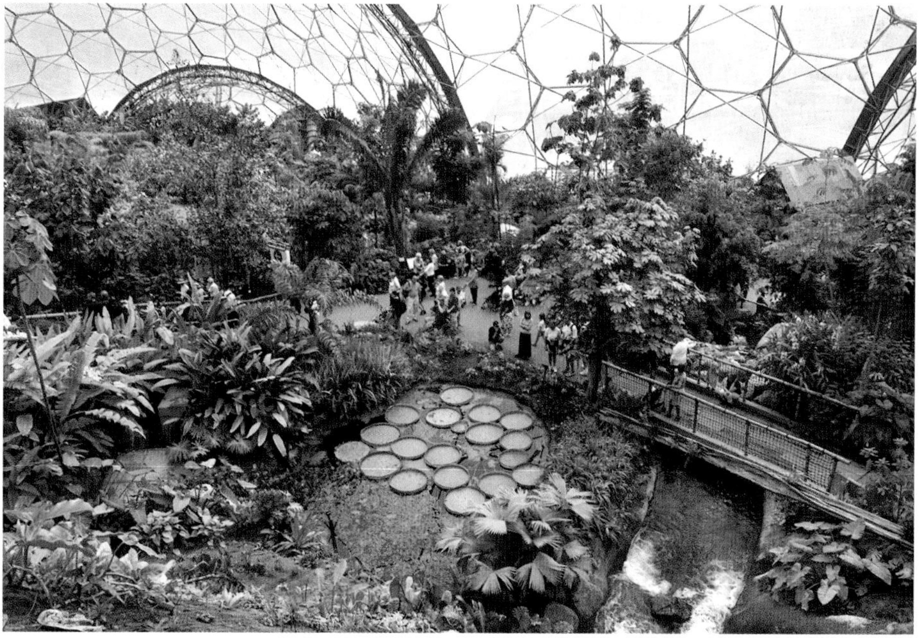

PLACES TO VISIT

BOTANIC GARDENS

Most botanic gardens have collections of ferns. Some that are particularly distinctive include:

Belfast Botanic Gardens, College Park, Botanic Avenue, Belfast, Northern Ireland BT9. Telephone: 028 9031 4762.
Website: www.belfastcity.gov.uk/parksandopenspaces
The interior of the 1889 Tropical Ravine house is designed as a sunken gully with a balcony from which the visitor can view the subtropical planting.

Benmore Botanic Garden, Dunoon, Argyll, Scotland PA23 8QU.
Telephone: 01369 706261.
Website: www.rbge.org.uk/the-gardens/benmore
Benmore is now a regional garden of the Royal Botanic Garden Edinburgh, and the huge free-standing fernery built in the 1870s was fully restored and restocked in 2008–9.

Birmingham Botanical Gardens, Westbourne Road, Edgbaston, Birmingham B15 3TR. Telephone: 0121 454 1860.
Website: www.birminghambotanicalgardens.org.uk
The Fern Walk was originally excavated between 1862–8. It was restored from 1996.

Chelsea Physic Garden, 66 Royal Hospital Road, Chelsea, London SW3 4HS.
Telephone: 020 7349 6458. Website: www.chelseaphysicgarden.co.uk
The Thomas Moore Fernery was built in 1907 on the site of the original of 1862–3 and contains a display of ferns described by him. There is also a reproduction travelling Wardian case on display.

Royal Botanic Garden Edinburgh, Inverleith Row, Edinburgh, Scotland EH3 5LR. Telephone 0131 552 7171. Website www.rbge.org.uk
The atmospheric 1960s Ferns & Fossils House contains a grove of tree ferns and other species planted around a cascading stream.

Glasgow Botanic Gardens, 730 Great Western Road, Glasgow, Scotland G12 0UE. Telephone 0141 276 1614.
Website: www.glasgow.gov.uk
Includes the wonderful Kibble Palace glasshouse, which re-opened in 2006 after a £7 million-restoration, replanted with exotic ferns.

Royal Botanic Gardens, Kew, Richmond, Surrey TW9 3AB. 24-hour Information Line: 020 8332 5655. Website: www.kew.org
Good fern collection in the glasshouses.

Southport Botanic Gardens, Bankfield Lane, Churchtown, Southport, Merseyside PR9 7NB.
Telephone: 01704 214 164. Website: www.visitsouthport.com
A unique glazed fernery set in classic Victorian botanic gardens, opened in 1876, restored and replanted in 2007–8.

GARDENS AND FERNERIES

Information on National Trust properties (marked NT) can be found at www.nationaltrust.org.uk

28 Kensington Road, St George, Bristol BS5 7NB. Telephone 0117 949 6788.
Website: www.victorianhousegarden.pwp.blueyonder.co.uk
Open for the National Gardens scheme, private visits STRICTLY BY APPOINTMENT ONLY. A classical 'ruin', statuary, grotto, pool, stumpery, artificial rockwork, tropical plants, tree ferns and native ferns all within a space 20 feet x 18 feet.

Ascog Hall, Ascog, Isle of Bute, Scotland PA20 9EU.
Telephone: 01700 504 555. Website: www.ascoghallfernery.co.uk
One of the best examples of a glazed outdoor fernery. Restored in 1997 and restocked according to the original 1879 planting plan.

Belsay Hall, Belsay, Northumberland NE20 0DX.
Telephone: 01661 881636. Website: www.english-heritage.org.uk
Impressive and dramatic quarry garden with many ferns.

Biddulph Grange (NT), Grange Road, Biddulph, Staffordshire ST8 7SD.
Telephone: 01782 522447.
The amazing gardens include the best-surviving Victorian stumpery in Britain. There is also a reproduction travelling Wardian case on display.

Birr Castle, Birr, County Offaly, Ireland. Website: www.birrcastle.com
A picturesque fern garden with streams and bridges, a fountain and moss-covered rocks.

Brantwood, Coniston, Cumbria LA21 8AD.
Telephone: 015394 41396. Website: www.brantwood.org.uk
Home of William Linton, author of *The Ferns of the English Lake Country*, and

subsequently John Ruskin. The new Fern Garden contains over 250 different types of British native ferns.

Brodsworth Hall, South Yorkshire DN5 7XJ.
Telephone: 01302 722598. Website: www.english-heritage.org.uk
15-acre garden dating from the 1860s, including a restored quarry garden planted with a fern collection in modern times.

Canonteign Falls, Christow, near Exeter, Devon EX6 7NT.
Telephone: 01647 252434. Website: www.canonteignfalls.co.uk
Probably England's highest Victorian fern garden, in a glorious setting.

Cornish gardens
The most important that contain tree ferns include: *The Lost Gardens of Heligan, Glendurgan* (NT), *Penjerrick, Trebah, Trewithen*, and *Caerhays*.

Trevarno, Crowntown, near Helston, Cornwall TR13 0RU.
Telephone: 01326 574274. Website: www.trevarno.co.uk
Tree ferns, plus the National Museum of Gardening, the largest and most comprehensive collection of garden tools and memorabilia in the country, including fern-related items.

Trewidden Garden, Buryas Bridge, Penzance, Cornwall TR20 8TT.
Telephone: 01736 351979. Website: www.trewiddengarden.co.uk
Tree Fern Pit in an old open-cast mine, which contains one of the largest collections of *Dicksonia antarctica* in the Northern Hemisphere.

Cragside (NT), Rothbury, Morpeth, Northumberland NE65 7PX.
Telephone: 01669 620333.
The roofless remains of a tropical fernery and temperate fernery, with low walls, rockwork constructed of large boulders of Fell Sandstone, pools and ferns.

Derreen Garden, Lauragh, County Kerry, Ireland.
Website: www.discoverireland.com
A 90-acre estate of bare rock was transformed into a luxuriant subtropical garden from the 1870s, including the great tree fern glen.

Dewstow Gardens, Caerwent, Monmouthshire, South Wales NP26 5AH.
Telephone: 01291 430444. Website: www.dewstow.com Probably the best Pulhamite garden in Britain, including large rock-edged pools, tunnels, and extensive subterranean tufa-lined grottoes. Includes a new stumpery.

Dyffryn Gardens, St Nicholas, Vale of Glamorgan, Wales CF5 6SU.
Telephone: 029 2059 3328. Website: www.dyffryngardens.org.uk
Edwardian gardens. An attractive outdoor fernery and stumpery with many well-labelled ferns, created in the twenty-first century in a quarry.

Eden Project, Bodelva, Cornwall PL24 2SG.
Telephone: 01726 811 911. Website: www.edenproject.com
As well as tree ferns outside, the huge rainforest biome contains a number of exotic ferns.

Fota House and Gardens, Fota Island, Carrigtwohill, County Cork, Ireland.
Website: www.fotahouse.com
An island garden of tender and exotic plants, including an outdoor fernery with large Tasmanian tree ferns.

Greenway (NT), Greenway Road, Galmpton, near Brixham, Devon TQ5 0ES. Infoline: 01803 842382.
Tree ferns in the garden running down to the River Dart. A rock garden was remodelled into a delightful fernery sometime between 1870 and 1882. Extensively replanted in 2003.

Hartland Abbey, Hartland, Bideford, North Devon EX39 6DT.
Telephone: 01237 441 264/234. Website: www.hartlandabbey.com
Small, outdoor, rockery type of fernery. Restored 1998.

Ickworth (NT), The Rotunda, Horringer, Bury St Edmunds, Suffolk IP29 5QE.
Telephone: 01284 735270.
Small, recreated stumpery in the garden.

Kingston Lacy (NT), Wimborne Minster, Dorset BH21 4EA.
Telephone: 01202 883402.
A very attractive and large outdoor fernery constructed in the late nineteenth century. Restored by the Trust and now extensively planted with over thirty-five varieties of hardy ferns, particularly male ferns.

Manderston, near Duns, Berwickshire, Scotland TD11 3PP.
Telephone: 01361 882636. Website: www.manderston.co.uk
One of the last great Edwardian houses to be built in Scotland. The range of glasshouses in the walled gardens includes a fernery.

Morris Arboretum, 100 E. Northwestern Avenue, Philadelphia, PA 19118, USA.
Website: www.business-services.upenn.edu/arboretum

Built for John Morris in the 1890s, the Dorrance H. Hamilton Fernery is the only remaining Victorian freestanding fernery in America. Restored in 1994.

Newstead Abbey Park, Nottingham, Nottinghamshire NG15 8NA. Telephone: 01623 455 900. Website: www.nottinghamcity.gov.uk/newstead
A small fern garden constructed from Derbyshire tufa in 1864–5, with rare terracotta fern bricks. There is also a rockery with a stream and a tunnel grotto.

Normanby Hall, Normanby, Scunthorpe, North Lincolnshire DN15 9HU. Telephone: 01724 720588. Website:
www.northlincs.gov.uk/NorthLincs/leisure/museums/normanbyhall
An example of a fernery as part of a range of glasshouses in a Victorian walled kitchen garden.

Overbeck's (NT), Sharpitor, Salcombe, Devon TQ8 8LW.
Telephone: 01548 842893.
A subtropical coastal garden. In the Edwardian house is a 'fern' fire screen, comprising a mixed collection of pressed ferns between glass plates set in a bamboo frame.

Peckover House (NT), North Brink, Wisbech, Cambridgeshire PE13 1JR.
Telephone: 01945 583463.
Peckover has a small twentieth-century, lean-to glazed fernery that in 2008 was awaiting restoration, and a small outdoor fern garden. In the house there is a good example of a 'fern' cast iron umbrella stand.

Rippon Lea Estate, 192 Hotham Street, Elsternwick, Victoria 3185, Australia.
Website: www.ripponleaestate.com.au
A vast fernery originally built in the 1870s and remodelled, *c.* 1884.
The largest extant shade house in Australia.

Sizergh Castle (NT), Sizergh, near Kendal, Cumbria LA8 8AE.
Telephone: 015395 60951.
The 1926 rock garden contains a fern collection started in the Victorian period, now one of the largest and most important in the UK, including the National Collections of *Asplenium scolopendium*, *Cystopteris*, *Dryopteris* and *Osmunda*.

The Swiss Garden, The Shuttleworth Collection, Shuttleworth (Old Warden) Aerodrome, near Biggleswade, Bedfordshire SG18 9EP.

Telephone: 01767 627927. Website: www.shuttleworth.org
The Swiss Garden (formerly Old Warden), contains a fantastic iron and glass cruciform-shaped Pulhamite grotto/fernery.

Tatton Park (NT), Knutsford, Cheshire WA16 6QN.
Telephone: 01625 374400. Infoline: 01625 374435.
Website: www.tattonpark.org.uk
One of the largest surviving private glazed ferneries in the country, designed by Joseph Paxton in 1859, and stocked with some of the original tree ferns.

MUSEUMS

Allhallows Museum of Lace and Local Antiquities, High Street, Honiton, Devon EX14 1PG.
Telephone: 01404 44966. Website: www.honitonmuseum.co.uk
Makers of the world-famous Honiton lace often used the fern as a motif.

Coalbrookdale Museum of Iron, Ironbridge Gorge, Shropshire TF8 7DQ.
Telephone: 01952 884391. Website:
www.ironbridge.org.uk/our_attractions/coalbrookdale_museum_of_iron
Coalbrookdale made cast iron garden benches, chairs and umbrella or stick stands with fern designs.

Linley Sambourne House, 18 Stafford Terrace, London W8.
Telephone 020 7602 3316.
Website: www.rbkc.gov.uk/linleysambournehouse
The largely intact Victorian interiors include a window garden on the first half-landing as well as two fern cases on the front elevation.

Oxford University Museum of Natural History, Parks Road, Oxford, OX1 3PW.
Telephone: 01865 272 950.
Website: www.oum.ox.ac.uk/index.htm
A great Victorian building with exquisite carvings of species of plants, including ferns, in the capitals and corbels of the main court.

INDEX

Page numbers in italic refer to illustrations